THE IMPROBABLE PRIMATE

Clive Finlayson is a noted expert on the Neanderthals and has been researching their final stand in Gibraltar. He is Director of the Gibraltar Museum and Director of the Institute of Life and Earth Sciences at the University of Gibraltar, having trained in Oxford as an evolutionary ecologist. He is a member of the Academia Europaea. His previous books include *Neanderthals and Modern Humans: An Ecological and Evolutionary Perspective* (CUP, 2004) and *The Humans Who Went Extinct* (OUP, 2009).

THE IMPROBABLE PRIMATE

HOW WATER SHAPED HUMAN EVOLUTION

CLIVE FINLAYSON

UNIVERSITY PRESS

OXFORD
UNIVERSITY PRESS

Great Clarendon Street, Oxford, OX2 6DP,
United Kingdom

Oxford University Press is a department of the University of Oxford.
It furthers the University's objective of excellence in research, scholarship,
and education by publishing worldwide. Oxford is a registered trade mark of
Oxford University Press in the UK and in certain other countries

First Edition published in 2014
First published in paperback 2016

Impression: 1

Published in the United States of America by Oxford University Press
198 Madison Avenue, New York, NY 10016, United States of America

British Library Cataloguing in Publication Data
Data available

Library of Congress Cataloging in Publication Data
Data available

ISBN 978–0–19–965879–4 (Hbk.)
ISBN 978–0–19–874389–7 (Pbk.)

Printed and bound in Great Britain by
Clays Ltd, St Ives plc

CONTENTS

LIST OF ILLUSTRATIONS

PREFACE

NEVER MORE THAN ONE SPECIES OF MAN...

In 1950, Ernst Mayr wrote about our species and our evolution in a Cold Spring Harbor symposium which was dedicated to the origin and evolution of man.[1] The symposia held at the Cold Spring Harbor Laboratory in New York since 1933 have debated major discoveries in biology and are highly regarded as landmark meetings. Mayr was one of the century's leading and highly respected evolutionary biologists and a key contributor to the modern evolutionary synthesis. In 1942 he had published a seminal book *Systematics and the Origin of Species from the Viewpoint of a Zoologist*. In it he featured his biological species concept, defining species as groups of organisms capable of freely interbreeding with each other and producing viable offspring in the wild. Mayr's paper at the Cold Spring Harbor symposium was therefore a must for contemporary students of human evolution. He made several very pointed remarks that are worth recalling over 60 years later. Mayr recognized that without fully appreciating the correct categorization of humans, we would be unable to understand our evolution: 'The whole problem of the origin of man depends, to a considerable extent, on the proper definition and evaluation of taxonomic categories.' He recognized that the arrival of the fully upright human marked a significant and unprecedented departure from anything that had come before and surmised that this arrival in what he called a different adaptive zone exposed humans to new selection pressures. This departure from all primate models that had preceded it, was so marked that it deserved a higher taxonomic category than that of species. With *Homo* came a new genus and, I would argue, a highly improbable primate. Mayr then went on to make a remark for which he has been criticized by palaeoanthropologists ever since. Mayr clearly stated that 'Indeed, all the now available evidence can be interpreted as indicating that, in spite of much geographical variation, never more than one species of man existed on the earth at any one time'. Mayr, I will argue, was right even though today many scholars of our evolution would disagree, still preferring to award species status to fossils of *Homo* based on morphological criteria.

The possible exception to Mayr's statement could be the Hobbit on Flores. Its small stature may have prevented interbreeding with other humans, purely because of physical limitations, but we cannot be certain of this. The application of the biological species concept to allopatric populations, those separated from each other by geographical or other barriers, has always been problematic because it is impossible to know whether those populations might be capable of interbreeding were these impediments removed. The Hobbit, isolated in a remote world almost taken out of a Jules Verne novel, is an example of an allopatric population whose taxonomic status is difficult to determine. Hobbit aside, what is clear now is that *Homo sapiens* was a polytypic species, that is, highly geographically variable but all individuals capable of reproducing with each other as Mayr recognized, but no different from many other similar examples from the natural world. But there is no evidence to suggest that these populations were ever distinct enough for interbreeding to have been prevented. The bastion of the palaeoanthropologists who supported the many species of Man, the Neanderthals, collapsed with the clear evidence that our own lineage interbred with theirs to a sufficient degree that the signal was retained in our genome. Then came the Denisovans, another ancient lineage now also shown to have exchanged genes with our own. If these populations were able to interbreed, and behave like one and the same species, after hundreds of thousands of years of isolation, then the question is resolved and Mayr shown to have been right.

Not everyone agrees. Some palaeoanthropologists maintain a multi-species view while accepting that there was interbreeding. They argue that hybridization can occur in the wild today between closely related species.[2] Mayr understood well that different species sometimes hybridize but they form stable hybrid zones which are confined geographically.[3] However, this is not what we observe with the Neanderthals, the Denisovans, and our ancestors. If the genetic signal has been retained right down to today, interbreeding would not have been an isolated affair.

Some notable palaeoanthropologists have followed Mayr in declaring *Homo sapiens* to have been a polytypic species throughout its nearly 2-million-year-old history. Emiliano Aguirre, the great scientist who discovered the spectacular site of Atapuerca was one and Milford Wolpoff in Michigan was another.[4] When I first entered this world of the study of our evolution, back in the early 1990s, the case seemed closed. The Neanderthals were considered a different species from us and so were a number of others, like *Homo erectus* and *Homo heidelbergensis*. Linked to these distinctions was the Out-of-Africa model. It explained the global expansion of our species from the African continent in relatively recent times, perhaps no more than 50 thousand years ago; our ancestors replaced all the archaic *Homo* species that they came across during the epic journey. It really was about how superior we were and how we had left no room for anyone else on the planet. We just had to be a different species.

How wrong we were. Time has shown that Mayr, Aguirre, and Wolpoff were probably right. The criticism that will be levelled against this viewpoint is that it is clear that the evolution of *Homo* was not linear and species branched off repeatedly. Lineages certainly branched off repeatedly, many more times than even the most ardent species-splitting palaeoanthropologist might admit; but they were lineages, not species. It may be argued that palaeoanthropologists are using the definition of species in a different way, following concepts of palaeontological species for example.[5] But surely the only species concept that is verifiable is the biological species concept, which requires good species not to interbreed regularly or produce viable offspring in the wild. One-offs or hybrid zones reinforce the message that evolving species may be linked by intermediates but two lineages that produced viable offspring in significant numbers would on this argument be part of the same species. Neanderthals and humans, on this basis, are the same species; and so, I will argue, was *erectus*, *heidelbergensis*, the Denisovans, and any other lineage we might care to compartmentalize. My argument in support of integrating *erectus* and *heidelbergensis* also under *Homo sapiens* follows since these forms were ancestral to the Neanderthals, humans, and the Denisovans; their morphological differences are the product of small samples and the discontinuity of the fossil record and mask continuous evolutionary trajectories over 2 million years. If the end products, after those 2 million years, were still behaving as one and the same species then it is unlikely that earlier forms, which were separated by less time, would have been any different.

In the early days I was taken by the multi-species idea and also by the Out-of-Africa replacement model. It seemed clear-cut and beyond dispute. But as I started to read more papers and books, and to think about all that I had been taught as a zoologist about the biogeography and evolution of species I began to question the widely accepted model. I was then only getting started in the world of human evolutionary studies and the views of a newcomer, and one who was not even a palaeoanthropologist, cut no ice. But the field of human evolution is first and foremost biology; the field of human geographical dispersals and extinctions is biogeography; and the field of the dynamics of human populations is population genetics. So let's use these tried and tested tools to understand ourselves. We might do better than developing hypotheses on the basis of metric measurements of isolated human crania when we cannot get a handle on the natural variation within the populations from which the crania came.

Once we emancipate ourselves from the shackles of rigid multi-species thinking the world opens itself up in front of us. Our evolution, from *erectus* on, concerned anatomical changes: those changes involved the tweaking of the starting body plan but they did not involve new designs. The earliest *erectus* differs from the most recent *sapiens* in degree. The process of evolution depended on each population; those in the arid areas of southern Middle Earth (Fig. 1; see also Chapter 5) kept on going down the lightweight and gracile line; those populations further north reinvested in muscle for power; and we do not know what *erectus* in south-east Asia did. Populations kept mixing and separating, generating the impression of a mosaic evolution of features. We should expect the greatest degree of contact and mixing to be precisely in the core area of Middle Earth, because it was an extended hub with similar climatic and environmental conditions throughout. No population was archaic, no population was modern: some were simply older and others more recent. Each was adapted to its own particular moment.

I will argue in this book that water was a key ingredient in shaping our evolution. The differences that we observe, for example, between the gracile kinds of humans and the more muscular ones, such as the Neanderthals, had a lot to do with water. The need to drink water daily unifies all humans, past and present ones. Yet most emphasis in human evolutionary studies is on food: what was eaten and how it was procured. Rarely does water enter the discussion. I will propose that the patchy distribution of water across the landscape in arid and semi-arid areas was critical to the origin and evolution of humans capable of traversing large tracts of land efficiently and speedily. Lightweight and gracile bodies behaved optimally under such conditions and muscularity was sacrificed in such situations. In addition to this, the problem-solving, information storage and retrieval abilities of good brains were particularly favoured in situations where choices of where and when to go for water determined whether individuals survived or died. Only in those parts of the world where water sources were common, could the investment in bulky, muscular bodies proceed unchecked. In the wet regions of Eurasia, close to the oceanic influence of the Atlantic and where topography accumulated water, the Neanderthals

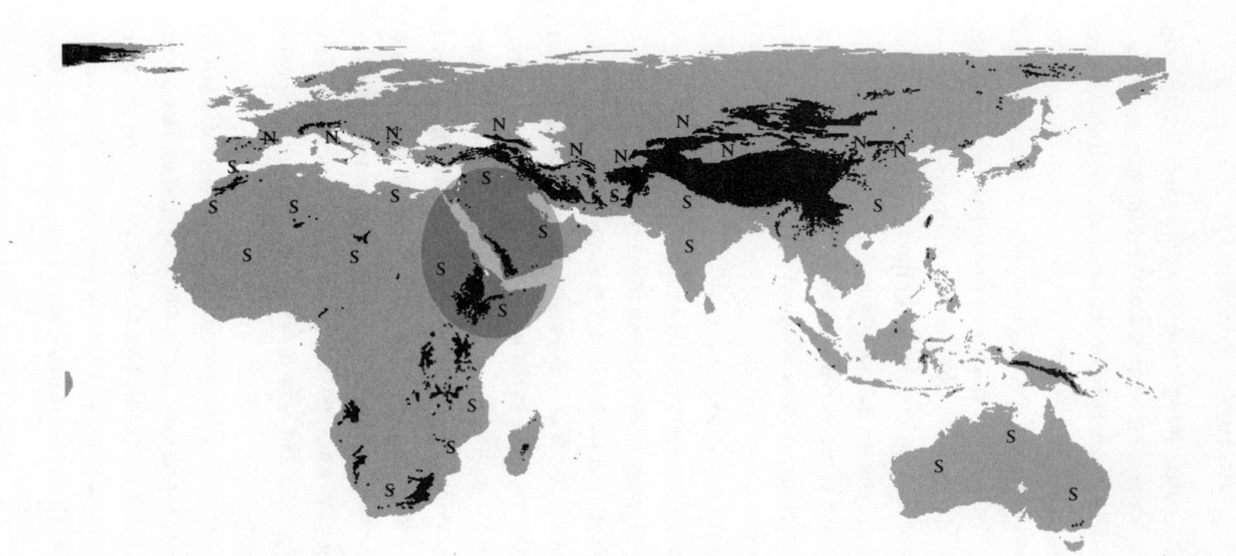

FIGURE 1. Map showing the geographical extent of Middle Earth. N = northern Middle Earth, a region that was colonized by humans from the core or southern Middle Earth; with a cool climate and short winter daylength this part of Middle Earth was always harder for humans to live in than southern Middle Earth. S = southern Middle Earth, a region of relatively continuous human presence, punctuated by periods when deserts or rainforests created barriers. Ellipse marks core Middle Earth which is defined in this book as the cauldron of human evolution.

epitomized this particular strategy. The differences between the extremes of a continuum—our lineage and the Neanderthals—were nevertheless insufficient to have produced two distinct biological species. They were, instead, geographical components of a widespread polytypic species. Seeing our evolution as that of an evolving polytypic species clarifies our behavioural evolution too. The use of different kinds of stone tools, the co-occurrence of distinct ways of making stone tools in the same site, and trends towards an increasingly lightweight and portable kit are best understood in the context of this evolving species, with its geographical variants, and with water as a major driver and marker of differences.

If taxonomic splitting of our species into distinct species has hindered how we see and understand ourselves, the typological classification of stone tools has set the understanding of our behavioural evolution back decades. It is time that we stopped talking of stone ages as if they were defining markers of our behavioural and cognitive evolution. There is too much overlap and variability for such a simple and clear-cut picture of Lower-Middle-Upper Stone Ages or Palaeolithic to make any sense. It follows that if stone ages did not exist, except in the minds of archaeologists, transitions between different stone ages are just as irrelevant. It gets worse. How important were stone tools in the daily lives of people and how defining were they of the people who made them?

The toolkit of the Mardu people of the deserts of Australia consists of multi-purpose appliances and instant tools (see Chapter 10). How much of that kit was stone? Not much. A few flakes and stone pounders and some stones used as anvils when grinding seeds. The rest was made out of wood or vegetal fibre.

The 780-thousand-year-old evidence from Gesher Benot Ya'aqov (GBY; Chapter 6) shows us how important wood and plant matter were to humans that far back. Wood and other plant matter only preserves under very special situations, such as the waterlogged conditions at GBY. Most of the time, such materials rot and disintegrate. So we really have no handle on the tool and weapon kit people had but the little we have hints at the importance of wood: 400-thousand-year-old wooden spears in Schöningen (Germany) or the rich array of wooden implements at the 40-thousand-year-old Neanderthal site of Abric Romani (Spain).[6] It makes sense that wood should have been the main material used by humans throughout our evolutionary history. After all, trees were always available in the habitats that we chose to live in. Just to show how this view impacts on some interpretations of material culture, the use of bone and ivory by the people of the Eurasian Plain after 45 thousand years ago has often been cited as an example of behavioural modernity when all it may show is the flexibility of humans in adapting to circumstances. There were few trees on the open steppe of the Eurasian Plain. Instead there were reindeer, providing antlers, and woolly mammoths, providing ivory and bone. People even made the superstructure of their tents out of mammoth bone.[7] This is not modernity, it is adaptability and improvisation. It had not happened earlier because nobody had entered the open steppe of Eurasia before that.

So what is modernity? I do not know. I would discard the concept altogether. Anatomical modernity is defined with hindsight, a poor way of doing science: we are the latest and so our anatomical features are supposed to be those that characterize being modern when what we really mean is most recent. The same criticism can be levelled on behavioural modernity. The problem once again lies with the application of the term. In the archaeological literature it has become linked to cognitive superiority. Humans became modern because they were cognitively superior to others, who by this definition were archaic. So the most recent (modern to the archaeologists) people on the planet are considered to be cognitively superior to all those who came before. In effect this puts us at the pinnacle of the evolutionary pyramid; all others, even the Neanderthals, must be at least one step below. If modernity, used in this sense, implies cognitive superiority, does this mean that we are cognitively superior to our parents and grandparents and their own parents who may not have had aeroplanes, fridges, or the Internet? Does it mean that 21st-century humans are cognitively superior to the Romans? Clearly they are not. Why then must the humans of 20 thousand years ago be cognitively superior (or modern) to the Neanderthals of 40 thousand years ago? Is it because they had different technology and did things differently?

Archaeologists have defined behavioural modernity ambiguously in relation to types of tools and other material culture. How do we distinguish, for example when we find a flint blade in an archaeological context, between the functional use of the blade and the evolutionary position of its makers? Simply put, was the blade made to fit a particular purpose or was it made because that was what that particular group of people did? It is impossible to distinguish between the two. The people of 45 thousand years ago were no more modern than their predecessors or their 'archaic' neighbours, just as we are no more modern than the people of 45 thousand years ago. We have confused the cumulative effects of culture in a social species with behavioural progression.

The thread of human evolution over 1.8 million years ago has therefore been one of adapting to an increasingly arid world while being tied down to the need to drink water regularly. Biologically, this has been expressed by the enhancement of those existing features that made this possible: bigger brains, lighter bodies, longer hind limbs. Behaviourally, it has been expressed through the development of an increasingly multi-purpose and lightweight kit. That is the thread but there were many variations and reversals along the way. The indigenous people of Australia or the westerners that first met them were no more modern than the first populations of *Homo sapiens*. Each did very well in the context of its time. The western settlers who made contact with the indigenous people of Australia made the basic mistake of confusing them for backward savages. I cannot help feeling that many contemporary archaeologists and palaeoanthropologists have made the same mistake when judging the peoples of the past.

As I was finishing this book, a paper was published[8] that reported an ancient African paternal lineage which had not been previously detected by geneticists. The time of the most recent common ancestor of this African-American Y chromosome lineage is estimated at 338 thousand years ago. The paper reported that the date exceeded the oldest anatomically modern human fossils by well over 100 thousand years ago. In palaeoanthropological language, the new date would put this ancestor within the realm of *Homo heidelbergensis* and not *sapiens*, a real problem of interpretation. I do not have such a difficulty. On the contrary it serves to confirm the unity of *Homo sapiens* and that the classification of Middle Pleistocene humans as a separate species—*Homo heidelbergensis*—was wrong.

As this book was going to print, the sensational fifth skull from Dmanisi, Georgia, was published in the journal *Science.*[9] The 1.8-million-year-old skull, attributed to *Homo erectus,* was significantly different in shape from the other skulls from the same site. Put together, the five skulls were as variable as African fossils that have been traditionally classified as three distinct species: *H. erectus, H. habilis,* and *H. rudolfensis.* The authors concluded that 1.8 million years ago there had been, in fact, only a single species of *Homo,* for them *erectus,* on the planet. This conclusion supports the idea put forward in this book that we can regard the observed variability of available specimens as representative of variation within a single biological species. If this was by now clear for the most recent branches of the human tree (*sapiens* and *neanderthalensis*), the latest findings show that it was also applicable to the earliest *Homo.*

The ideas presented in this book are my own but they have benefited from discussion with many colleagues and friends. I am most grateful to all of them. My wife, friend, and colleague, Geraldine, has been my prime assessor; she has always had time to discuss and debate an insight, an idea, or a comment, often at the drop of a hat. This might have been while excavating inside a cave, while driving on a motorway, or simply over dinner. She has kept me on track too, warning me of the pitfalls of some of my arguments. Her unique knowledge of habitat structure has been critical for this book.

My son Stewart is my natural history companion, sharing many hours in remote locations with me. His fresh views on animal behaviour and ecology and his passion for caves have kept my own spirit alive. He understands the importance of a good grounding in natural history before even hoping to try to relate to those great naturalists who were our ancestors, often commenting with incredulity on unrealistic remarks that he has read in scientific papers.

Darren Fa, my former PhD student, is a friend and colleague who has also been deeply involved in our research programme for a long time and who, with Geraldine, participated in an early presentation of our understanding of human biogeography, back in 2000. As a marine biologist, his knowledge and insights of human activity along the intertidal zone are making an important contribution to this field.

I would like to thank all friends and colleagues who have worked with me in the study of our origins and have in some way contributed to the ideas put forward in this book: Joaquín Rodríguez Vidal, Francisco Giles Pacheco, Larry Sawchuk, José Carrión, Juan José Negro, Richard Jennings, Jordi Rosell, Ruth Blasco, Marcia Ponce de León, Christoph Zollikofer, José María Gutierrez López, Alex Menez, and Antonio Sanchez Marco.

Finally, I thank Latha Menon and Emma Ma at OUP for their support, patience, professionalism, and hard work in the preparation of this book.

1

The Inverted Panda

If we were able to go back to the Middle Miocene world of 16 million years ago, when apes were widespread across large areas of the Old World,[1] we would be forgiven for not predicting the future existence of a creature that would one day call itself *Homo sapiens.* We might have predicted something like a gorilla, an orang-utan, or a chimpanzee but not a human. Yet this improbable primate's heritage is in the deep forests of the Miocene apes and it is here where we should start looking for the antecedents to the path that led to humans.

Let us start with the brain. There is, after all, no other organ that defines us better. Katherine Milton at Berkeley gave us a great and convincing insight into the function of the primate brain in a forest context.[2] She highlighted the complexity of the rainforest world that the early insect-eating ancestors of the primates entered at the end of the Cretaceous, some 70 million years ago. This was a world that was coming under the dominance of the flowering plants and some of these early insectivorous mammals were probably drawn up into the trees where insects gathered round flowers. Once up there, flowers and young leaves may have been added to the diet. We can imagine a scene with shrew-like mammals scurrying among the branches of an ancient forest, snapping away at juicy insects clustered around ancient flowers. Occasionally one might take a bite at an insect and accidentally swallow a petal. If it liked the taste, and there was no ill effect, petals might have been added to this individual's diet. Let us imagine that these animals with a wide tolerance in the recognition of insects, which may have allowed for such mistakes, could have gained some advantage from consuming petals. In time the forest might have been swamped with petal-snapping descendants and any genetic novelty improving petal-snapping would have been favoured. I shall say more about how behaviour can predispose animals to particular genetic novelties in Chapter 4. The traits that characterized the primates thus evolved as plants became increasingly important in the diet, the grasping hand being an early innovation that has stayed with us until today.

Once in the difficult, three-dimensional forest canopy any improvements to the visual apparatus would have been favoured. This is because visual discrimination—colour vision, sharpened acuity, and depth perception—would have helped to detect ripe fruits and tiny, young, and succulent leaves from among the vast expanses of green. The green world of the forest canopy may seem luxuriant to us viewing it from the outside but this is a false impression. It is not an easy world at all for a primate. High-quality foods, particularly fruit, are distributed in patches with lots of unsuitable trees in between. These foods are often seasonal so a knowledge of when, as well as where, to locate the good patches is critical to survival. The canopy actually resembles an open ocean or a plains environment in the sense that life is boom or bust: find a good patch at the right time of year and you will flourish but the consequences of not finding such bonanzas can be fatal.

Forest primates face two dietary choices: eat large quantities of low-quality food, such as leaves, which are widespread and abundant but low in energy and high in fibre content or, instead, focus on the high-quality foods, such as fruit, which are rich in carbohydrate and relatively low in fibre but which require time and effort to find. Those eating fruit need protein supplements which they often obtain by eating insects or other animal matter. Primates that have taken the abundant, low-quality food route have evolved specialized guts that allow them to process bulk and extract as much energy as possible from the mature leaves which they eat. In general they tend to be relatively inactive, to conserve energy, in comparison with those primates that take the high-energy foods. Think of a group of gorillas and you will get the picture. The highly specialized giant pandas of China or the tree sloths of central and South America, mammals unrelated to the primates, paint a similar picture.

Those that have gone for the scarcer and patchier high-quality foods have, instead, relied on behaviour, especially a good memory that allows them to remember where and when to go for the choice items. Among these primates we could include the spider monkeys of Panama[3] and chimpanzees.[4] Improved visual and cognitive skills gave direct foraging benefits to the early primates, promoting large brains—the hallmark of primates from the very beginning. That brain was put to good use by a wide range of primates, in particular those which went down the path of feeding on high-energy, difficult-to-find foods. So our brain is also probably an outcome of the fruit- and foliage-seeking primates' behaviour, tweaked and enhanced along the 70-million-year journey to the present. The brain expansion that our ancestors experienced over the last 2 million years as they left the forests would not have happened without the head start that the forest primate brain gave them.

One of my earliest recollections of watching birds in an English forest, when I was a student in Oxford, is of the mixed winter flocks of tits and other insect-seeking birds in the nearby Wytham Woods. I would spend time searching what appeared to be an empty wood only to suddenly hear the call of a blue tit or a goldcrest. Soon I would be overrun by a noisy cacophony as several species of birds swept past me at canopy level, taking what they could find on the way. At the scale of the insects which were their prey, I can only imagine that this innocent-looking and quaint flock of colourful birds would have been equivalent to a mixed pack of hyenas, lions, and cheetahs flushing prey across a savannah. Many years later I observed the same winter-bird phenomenon in a mountain forest in southern Spain and a similar concert of nomads among desert birds on the island of Fuerteventura.

For a number of years now I have been lucky to go out to sea to observe marine birds in the Strait of Gibraltar. September is an excellent time because it is when the flying fish are migrating. You can spend hours searching the horizon for birds and you can travel many miles of empty sea but from time to time you hit a feeding frenzy. A pod of dolphins has found a shoal of flying fish and they are after them. The fish make their escape by darting out of the water and gliding to safety. But the seabirds that have been absent in our journey suddenly appear from all directions. Shearwaters, gannets, gulls are all around us, chasing the fish on the

FIGURE 2. Griffon vultures feasting on a carcass, typifying the boom-or-bust world of species, including humans, dependent on patchily distributed resources in space and time.

surface or diving after them. In a matter of minutes hundreds, even thousands, of birds are all around us making the most of the bounty (Fig. 2).

I have also spent many days in a remote mountain in the Pyrenees, sitting inside a cold wooden hide waiting for vultures to come to a carcass. Sometimes it is a dead deer, at others a goat or a pig, and they have been left there as bait by park wardens. My aim has been to photograph the vultures. Here we can get all four European species together including the now rare lammergeyer—the 'bone breaker'. I have spent days seeing nothing; staring at an empty place, desolate and freezing. But it has been worth it when the birds have come. As with the tits and the shearwaters you go from nothing to a frenzied mass of vultures in minutes. Once an entire deer was reduced to a skeleton in 17 minutes and then the birds were gone once more, leaving the cleaned bones as souvenirs.

These apparently disparate examples—tits, dolphins and shearwaters, and vultures—have one thing in common. At the scale at which these different predators are operating they perceive their world as patchy—they are worlds of needles in haystacks. The needles are rich gifts—insects, flying fish, carcasses, trees in fruit. When found they provide much-needed energy that permits continued existence. Many primates also exploit patchily distributed resources and many of them move around in groups. Groups hold the all-important key to finding the coveted metaphorical needles. This key comes with an expensive price tag: cooperation.

In an ideal world, an individual predator would want to take as much energy (food) out of its environment as it can, without having to share it with others. But in many situations an otherwise selfish predator is obliged to cooperate—better to find the cake and take a slice than risk not finding the cake at all. A number of factors will influence the size of a group and the degree to which its members may be prepared to defend a patch of food. The size of a feeding group will depend on the size of the patch of food. This is a simple question of physical limitation—the smaller a patch, the fewer the number of animals that can feed in it. If an area has patches of different sizes, groups may fuse or split into smaller units to take the size of patches into account. This adjustment of feeding-group size is typical of many primates. The density of food inside a patch will also determine how long a group can stay feeding in it. If the density is high enough to keep them there for a while, then the patch may even be worth defending against other groups and individuals within the group may themselves jostle for the best positions within the patch.[5]

There is one other—all-important—ingredient that I want to add to this soup. That is the actual number of patches within a group's home range and how that density changes as time passes. An area may have lots of good patches in summer and fewer in winter or the number of these may change from one summer to the next, and so on. All this translates into how easy or difficult it might be to predict and find good sources of energy or other essential commodities. When we come to look at how our ancestors became increasingly terrestrial in habits and spent more and more time in open habitats, we will realize how important the density of good patches of food would have been. Importantly for the thread that I will develop, the density of water patches in an otherwise arid landscape was a vital ingredient of our story. For now, let us understand how the number of patches within an area—their density—can affect the behaviour of primates.

In primates there is a strong relationship between the amount of fruit eaten and the size of a group's home range. This is because quality fruit patches tend to be very scattered so that primates feeding on fruit have to travel greater distances to secure food than others which feed on easier-to-find foods such as leaves. So it follows that these fruit-eaters spend considerable amounts of time and energy moving about. These patches are high premium and will be defended when other groups are encountered but very often rival groups are so thin on the ground that encounters may be quite rare. It is only when an area holds a high density of animals that aggressive interactions between groups may become a fact of life. When thinly spread on the ground, groups may still want to defend good patches but they may not be able to do so because all the trees within their large home ranges cannot be efficiently monitored at once. The idyllic image of a beautifully balanced Nature masks the underlying reality. Animals constantly face uncertainty and their environment continuously puts them to the test. Humans have been a part of this uncertain world even though we have tried hard to tame it. For now, let us stay with the thought that living in groups is another consequence, like large brains and grasping hands, of a fruit-eating life in the forests.

Living in groups, in regular contact with family, friends, and competitors, poses its own challenges. For each individual, the other members of the same group are part of its environment. If the outside world is complicated, imagine how much more tortuous it becomes when some components of that environment are like you and are constantly changing their minds or are trying to change yours. The game becomes really convoluted in this fluid world. A big brain comes in handy in these circumstances but the problem is that your friends and foes also have large brains. If investment in a life of eating high-quality, highly-dispersed foods promoted large brains and living in groups, then the consequent social life drove primates that adopted this way of life into an arms race. Bigger brains meant better interpreters and manipulators of society, which in turn encouraged even bigger brains.

All that I have described so far would have happened within the forest. Nothing required a primate to take the first step away from trees, not even to come down from the canopy onto the ground. But we do know that one day a primate did come to ground and eventually stepped away, perhaps nervously at first, from the forest edge. There are a number of views, some strongly conflicting, about how and when all this happened. I am not going to go into that now but we will do so in later chapters. What I want to highlight here is that something happened away from the forest. It may have been a development of something that the forest ancestors had been doing already or it could have truly been a novelty. Either way, it was taken to a new level: it was the inclusion of meat in the diet.

I realize that my last sentence may have sounded somewhat banal but it is very important to our story. In later chapters we will consider how meat-eating took off; if it was gradual, sudden, if meat was scavenged or hunted, if it was a small part of an omnivorous diet or if it was, instead, a major component. That does not matter right now. What matters is that we ended up with a primate walking on two legs and regularly eating meat in an open savannah. Such a primate entered a world that had until then been the exclusive domain of the carnivores—the big cats, dogs, and the hyenas. These predators had an ancient legacy of meat-eating, an inheritance that took them back tens of millions of years to the Late Eocene and Early Oligocene.[6] Our ancestors arrived very late on the savannah scene and they brought with them a vegetarian and arboreal heritage. They were interlopers, amateurs in the world of the professional carnivores. So it is quite amazing that they managed to eke out a living in this challenging environment.

We may learn something about how and why they became successful by comparison with other species. Can we find a species that, like our ancestor, has departed significantly from the lifestyle of its lineage? Well, it may seem surprising that I should pick the giant panda of China as a good candidate but hear me out. The specialization of the giant panda in eating bamboo is well established and it seems to derive from less-specialized pandas of the Pliocene, providing a unique example of the adaptation of a meat-eating lineage for herbivory.[7] So we have in the giant panda almost a mirror image of ourselves. We moved from a largely plant-eating line in the direction of meat-eating and the giant panda went the opposite way. You could say that we are inverted pandas. We both represent examples of species that have somehow managed to change their ways radically from the remainder of their relatives. Now, the giant panda succeeded in the specialized world of herbivores by focusing on bamboo, by becoming highly specialized in a trade that seems to have been largely neglected by other plant-eaters. But surely this cannot apply to humans? We are, after all, the epitome of the generalist. I take a different view which will become clear in the chapters that follow.

An adult female sits down to rest. She is tired and takes a bite from a ripe fruit while her restless children drink some water and carry on playing around her. Her partner steps away surreptitiously and tries to blend into the background. He is tired of all the start-stop movement involved in gathering. He has kept a safe distance as the female has jostled for position with other females trying to secure the best items. The highlight of his day was meeting a male friend who in the past has been with him and other males out on 'banner-waving' expeditions proclaiming territory and tribal identity. Soon the family party will be on the move once again as it prepares to head for the safety of its abode. Other females and juveniles meet up as they get closer to home, exchanging news and gossiping. There are males too and they slow down their pace to meet up with the other males, gossiping too about different kinds of things, all the time surveying the surroundings and keeping an eye on the youngsters and the females. Once in their core area the families split up to roost in their separate enclaves. Night is the time to sleep in the safety of your home.

Have I described a scene of a group of fruit-eating primates in a remote Miocene forest or does it sound like something closer to home? There is much in our story that is the result of evolutionary pressures, and much about the way we are and behave today, under the cloak of self-proclaimed civilized behaviour, that harks back to those very forest origins. Culture has cast a convincing veneer over our biology. But if we are able to shake it off we will find, underneath, an improbable primate, still behaving like an inverted panda.

2

And the World Changed Forever

7 TO 4.32 MILLION YEARS AGO

Many fruit-eating monkeys and apes will eat insects or other sources of protein in order to supplement their diet.[1] Dependence on a variety of foods would have started off as a way of balancing the fruit-based diet. Which foods to add and which to reject must have depended on what was readily available and on the willingness of the forager to take a chance at trying something new. In this risky world of opportunists some may well have fallen by the wayside when eating what they shouldn't have and, no doubt, the bright sparks in the group would have soon seen the effects and avoided the danger. That would have been one of many advantages of living in a group. Experienced mothers, just like chimpanzees, would have taught their young what to eat, how to eat it, and what to avoid too. In this way cultures would have spread within populations, as happens among chimpanzees today,[2] simply as individuals watched and learnt from one another. It could not have happened in solitary animals with small brains. Large brains and social life predisposed primates for culture at a very early stage. It is important that we appreciate that our remote ancestors of 7–6 million years ago,[3] the starting point of this chapter, would have already held the potential for cultural development—the legacy of life in the forest canopy.

But by this time it seems that these remote ancestors were no longer living deep in the forest canopy. We are in a time of significant change and upheaval: deserts begin to appear and open, grassy environments dominate over large areas of tropical Africa where our ancestors are living.[4] Nobody disputes that these dramatic climatic and environmental changes happened but there is considerable debate regarding the habitats that our ancestors actually occupied at this time. Knowing where our ancestors lived is as critical as knowing what they ate so it is important that we disentangle the debate. This revolves around whether our early ancestors lived in grasslands—'the Savannah'—or instead remained within largely wooded habitats, never away from trees.[5] A popular controversy also involves this period of our evolution although it is not usually entertained in academic circles.[6] I am referring to the Aquatic Ape Hypothesis which was first put forward by Sir Alistair Hardy in 1960[7] and has been championed since then by Elaine Morgan in particular.[8] So it is critical to our story that we take a look at the evidence and clarify the situation, including this long-standing dispute regarding our putative aquatic origins. To do so, I will rely on the available evidence, which concerns Toumaï (*Sahelanthropus tchadensis*) and Ramidus (*Ardipithecus ramidus*).[9]

I will start with Toumaï, which is the older of the two species. Its environment has been reconstructed using a combination of geological and fossil evidence.[10] Fossils can be good indicators of environmental conditions when we compare them to their closest living relatives. In some cases we may be looking at species which are still around today and we assume that their requirements have remained largely unchanged in the interval between then and now. I don't see this as a problem, especially in cases when anatomy is a good indicator of a species' habits: for example, a stork has exceptionally long legs for wading in shallow water and most of the world's storks live close to water, so finding a fossil stork strongly suggests that there was water somewhere close to where it died. When several species that are associated with shallow water are found fossilized in the same site the interpretation is strengthened—it is robust and gives us confidence.

Toumaï, at first hand, would not seem to be much help to us in clarifying matters. The geology of the deposits in which its fossils, along with other animals, were found is a kind of cemented sandstone which was very informative. It was a mix of wind-blown sand, from the nearby desert, and grains that had been reworked by Nature in a lake environment. These lake-deposit sand grains formed a matrix with mudstone and fossilized single-cell algae (diatomites) that would have lived in the ancient lake. The preserved ripple marks left by currents and waves in the lake along with the remains of sand bars gave an idea of the water movements that had taken place: these currents had not travelled in a single direction, as we would expect if a river had been flowing, but went in all directions. The conclusion was that the evidence indicated a place that had been flooded from time to time, draining at other times into the nearby desert environment. So the shoreline of the lake changed repeatedly in a battle with the ever-encroaching desert. This must have been an ever-changing and unpredictable landscape quite unlike anything within the forest canopy.

Putting this geological evidence together with the fossil fauna that was excavated alongside Toumaï's bones, the scientists working at Toumaï's site pieced together a varied world that existed between lake and desert. There were, undoubtedly, a variety of freshwater habitats judging from the richness of the aquatic and amphibious fauna. They found ten different types of freshwater fish belonging to seven distinct families, all but one having living representatives on Lake Chad today. One species—the tiger fish *Hydrocynus*—would have been a strict hunter of other fish that took its prey by sight in deep waters that were rich in oxygen. The water bodies must have been large as many of these predatory fish reached lengths of over 1 metre. This idea was supported by the presence of large numbers of fish-eating crocodiles that included a new species of gavial. Second, there would have been swamp vegetation, clearly indicated by fish that would have lived in turbid, oxygen-deficient waters. One species—a knife fish *Gymnarchus*—would have used an electric sensory system to find its way around and hunt in murky waters.

Many fossils reflect life on the sandy banks at the lake edge. The animals that would have frequented the lake shores were anthracotheres, extinct relatives of hippos, and hippos themselves. Even complete skeletons of hippos have been found fossilized. There were otters, softshell turtles, and a python too. Remains of liana-like papilionoid plants indicate the likelihood of gallery forest that would have fringed the lake. Colobine monkeys would have lived here too. Further away from the shoreline there would have been wooded savannah, indicated by the abundant presence of elephants and their extinct relatives the gomphotheres, as well as now-extinct giraffes.[11] Antelopes such as kob and their relatives indicate the presence of grassland where they would have grazed. The picture is completed by animals that probably moved between some of these habitats: land tortoises, monitor lizards, rodents, pigs, hyenas, and sabre-toothed cats.

This is quite a detailed picture of the location where Toumaï lived so why did I say, when I started this description, that it would probably not clarify matters regarding the environment exploited by our earliest ancestors? The reason is that we are unable to say whether Toumaï was aquatic, lived in the gallery forest with the monkeys, or ventured into the wooded savannah or even the grasslands. We know where many of the other species were likely to have lived because they left living relatives, because we have many fossils of extinct species, or because their anatomy is revealing. Not Toumaï, of whom we have a skull, a mandible, and not much else.

If Toumaï cannot resolve the question of the habitat which our earliest ancestors occupied, can Ramidus? Ramidus had been first described in 1994, from teeth and jaw fragments, but by the middle of the first decade of the 21st century the world awaited the publication of a much more complete set of specimens. In October 2009, *Science* dedicated an issue to a study of the remains of *Ardipithecus ramidus*, an unprecedented step for this journal. Eleven papers in all, published by a large international team, described this species and its habitat in great detail, based on 110 specimens, representing a minimum of 36 individuals recovered from 4.4-million-year-old levels in the Aramis region of Ethiopia. What was Ramidus like? A partial skeleton of a female allowed the researchers to estimate that she weighed 50 kilos and stood 1.2 metres tall. She had a brain the size of a chimpanzee's. Males were probably similar in size as little evidence of sexual dimorphism was found in this species. Examination of the teeth suggested that Ramidus was more omnivorous than chimpanzees and probably fed on the ground as well as on the trees. The front and hind limbs, as well as the pelvis, tell us that Ramidus moved ably on trees supported on its palms and feet but it was not specialized for a clinging or climbing tree life. It could also walk on its hind limbs when on the ground but it seems to have been less committed to the terrestrial way of life than the later australopithecines.

This is helpful. We know that Ramidus, as probably Toumaï too, was eating a wide range of foods and had become less dependent on fruit-eating than chimpanzees. A broadening of the diet would have had an impact on Ramidus' behaviour but what was its habitat? In spite of quite detailed information about the animals found in the same levels as Ramidus and also of the carbon isotope composition in teeth,[12] controversy broke out regarding the interpretation of the evidence. On one side were the researchers that published the 11 papers, headed by palaeoanthropologist Tim White in Berkeley, and on the other was a group headed by geochemist Thure Cerling at the University of Utah. White is a leading palaeoanthropologist who has spent much of his career working in African sites while Cerling is the leading light in the study of the emergence of C_4 grasses during the Miocene and Pliocene. So this was a debate between heavyweights and it centred on whether Ramidus lived in woodland (White's group) or more open grassland habitats (Cerling's group).

White's group collected fossils of large mammals along a 9-kilometre-long transect within the 4.4-million-year-old horizon that contained the Ramidus fossils. They collected an impressive

4,000 specimens which represented 40 distinct species.[13] Their interpretation of this fauna—dominated by leaf-browsers and fruit-eaters—was that it emitted a clear signal that the area where Ramidus and all these other species had lived had been dominated by woodland and not grassland. Only three primate species were recovered from the palaeontological beds and that included Ramidus which was the rarest of the three. In contrast, around a third of the large mammal remains came from a colobine monkey which lived in dense to open forest.[14] White and colleagues claimed further support for the woodland setting from the other numerically-dominant large mammal of the assemblage—the spiny-horned antelope which was a leaf-browser that is thought to have lived in bushy to wooded habitats.[15] In contrast, grazing antelopes were rare. The emergent picture of Ramidus' habitat, according to White's group, was of a dry woodland setting, with small patches of forest, far away from rivers.

Cerling's team focused on the interpretation of soil and teeth-isotope analysis and claimed that the results indicated a very different picture from that suggested by White's team. They weren't convinced by the fauna either. The antelopes could have been inhabiting thickets close to rivers and they argued that the absence of duikers meant that there was no dense woodland or forest.[16] Their conclusion: Ramidus lived in riverside forest or woodland in an otherwise open bushland and woody grassland.[17] So, while White placed Ramidus in a dry and fairly open setting, Cerling included habitats which were more closed and near fresh water.

Rather than focus on the differences, I would like to establish what there is in common between these apparently disparate views. First, both interpretations recognize the presence of aquatic species, which indicates that there must have been a river or wetland nearby. White's team places the river away from where Ramidus was living but does not discard the possibility, to my knowledge, that they were out of reach of water, especially when rivers overflowed during the wet season. The fossil-rich sediments were silts and clays that had been formed from materials deposited by rivers and lakes or on floodplains, and catfish, presumed to have been deposited during overbank flooding, were common. Hippos, crocodiles, freshwater turtles, and waterfowl completed the picture of shallow, seasonal, waters away from the main river.[18] This portrait is not dissimilar to that which we found for the proximities of Toumaï's territory.

I am very familiar with this world of seasonal flooding as I have spent many days of field research in the Doñana National Park in south-western Spain. Doñana is a huge area of marshland and dunes. It is hard to imagine how much life such an area can hold. The park had been a traditional hunting ground of kings who were obviously very aware of the rich game that lived within. For me the richest areas inside the park were always those in a part known as 'La Vera'. This is where all the rainwater that has percolated into the sand dunes and down to the water table seeps back out onto the surface. On the dry side of La Vera are extensive areas of aromatic shrubs—lavender, rosemary, rock roses—and scattered cork oak trees. On the wet side lies impenetrable marshland. This inland coastline, which is after all what La Vera is, changes shape with the season and from one year to the next. It all depends on how much rain fell in the winter. Some years, the marsh can become an inland sea. This happened in the winter of 1995–6, when huge amounts of rain fell. Every animal in the marsh sought refuge on dry ground. Some became trapped on sandy islands and died, many drowned, and the remainder were tightly packed in La Vera.

In dry winters, such as 1994–5, almost no rain falls. The marsh remains parched, which is its typical aspect during the summer months. We could drive over vast areas where ducks had swum in previous winters. Now they were littered with the carcasses of dead animals that had been unable to find water to drink. The following winter, others would drown in the very same places. In these times of drought La Vera remained the best place to find animals, as this was where they had the greatest chance of finding water, often by digging. The point I am making is that these areas that were subject to constant flux—winter flooding, a summer dry season, years of extremely high rainfall alongside years of almost no rain—attract many animals. They are a source of diversity and biomass.

We found a similar picture around shallow lakes and ponds that were scattered within the dunes themselves. They too had a shoreline that receded and expanded with rainfall and the response of animals was fine-tuned. In years when the water levels were not too high or too low some of the lakes were pink as thousands of flamingos poured in from distant lands to make the most of the optimum conditions. But other animals were not so fussy and they exploited the lake shores, and indeed La Vera, fully by being opportunistic. The wild boar comes to mind as the star of these amphibious lands. These intelligent and resourceful mammals were a permanent fixture and they left their mark as they dug deep into the mud for roots and invertebrates, transforming patches of ground in the process. But in this world of constant change the best tactic was to take everything that presented itself. One of the most amazing discoveries was that wild boar behaved like hyenas, seeking out the carcasses of dead deer and other large mammals, eating the intestines and the flesh and even taking the large bones away, cracking them open with their powerful jaws and extracting the marrow.[19]

So what is our take-home message so far? It is that areas subject to seasonal or between-year flooding may be in constant flux but they are also magnets for animals. When excessive rainfall floods low-lying areas, animals congregate on the marginal dry ground and when drought hits they gather around shrinking waterholes. Predators have plenty of scope in this world, and entrepreneurial species, like the wild boar, that are not too choosy about what they eat will find many and varied opportunities that will allow them to subsist in spite of changing conditions and without need to move very far. Our earliest cousins—Toumaï and Ramidus—probably lived in such a world or very close to it. So coming down to water, within the relative safety of trees at first, exposed these creatures to other sources of food that could be found and easily picked from the water's edge. Quite by chance, as with so many elements of our story, this new behaviour brought them close to one of the world's richest environments. It would have been a short step to get them into the water, wading, even swimming, after succulent morsels.

For a primate, the transition from living in the tropical rainforest to living in a tropical seasonal forest would have been enhanced by having swamps, marshes, and other kinds of wetlands, which behaved as I have just described, in close proximity. This is because these wetlands are among the most highly productive environments on the planet, more so than tropical seasonal forest and only slightly less productive than the rainforest itself.[20] Any primate faced with a shrinking rainforest and forced to live on the edge would have been better off with wetlands close by than it would have otherwise been if all it had around it was seasonal forest, woodland, and savannah.

If we return to the habitat descriptions that have been offered for Toumaï and Ramidus, we can find features in common that take us further in painting the picture of their home. All descriptions incorporate trees and the arguments fall around whether they were few and far between with plenty of grassy, open spaces (as in a wooded savannah) or whether they were, instead, tightly clumped (as in woodland or forest). There are so many possible permutations that we can make from the available descriptions that they tend to render the arguments futile, particularly as the spatial distribution of the habitats is unknown. You could easily have a home range of a troop of Ramidus that would have encompassed patches of forest, savannah, and wetland, and all could have formed part of the habitat of that troop with different amounts of time being spent in each. Then there are elements that are missing from the descriptions, most notably rocky outcrops.

The problem with habitat descriptions is that they tend to fall into the trap of allocating observations into boxes of predetermined habitat types. Much of the discussion between Cerling's and White's groups, for example, has had to do with how they have defined habitats. They have used UNESCO definitions[21] which may make some sense in today's world (although I even have my doubts about that) but which we really cannot use with the limited fossil data at our disposal. Perhaps then a better approach would be to define habitat from the perspective of the users, in this case the early hominids. We can do this quite easily by listing the elements of that habitat that we know were present at the sites. Those elements in the present example are seasonal

FIGURE 3. The human habitat: trees/open-spaces/water.

shallow water, trees, and open (treeless) spaces (Fig. 3). These were the essentials, it would seem, of the early hominid real estate. I could describe this as a mosaic but I would be falling into the predefinition trap myself. A mosaic of habitats only exists in our minds when we predefine habitats. If we find several of the predefined habitats in an area then we conclude that the species in question lived in a mosaic. But if we just stick to established elements, the variety of elements is instead what the species needed to make a home, its habitat.

Trees seem to have been a critical component of the habitat of these early hominids and this is hardly surprising given a heritage of living in the forest canopy. What we do not know is whether or not they spent any of their time in the canopy, whether they fed on the trees or whether, instead, they simply used them as places to which to climb for safety. White's study of the anatomy of

Ramidus concluded that it walked on two feet[22] but also spent time on the trees. Ramidus had not therefore committed to walking upright on the ground all the time. Such facultative bipedalism makes sense in a habitat in which trees and open spaces are interspersed, as seems to have been the case for Ramidus and Toumaï. Part of the time could be spent on the ground, in the open spaces or below trees, and movement on the ground may have been on two or four limbs. When on two limbs, the freed hands could have been used to collect food from shrubs or the lower branches of trees just as their ancestors would have done when walking on branches.[23] Hence the habit and accompanying anatomy would have been another benefit of adaptations arising from a life spent handling fruit in the canopy.

The other component seems to be shallow water, ephemeral in nature. This is another first for our lineage and we do not know to what degree they exploited this element of their habitat. We have to guess but we can do so by seeing what other primates do in such situations. If the lineages of these primates are older than the early hominids, we can infer that the latter would probably have had the capability to perform behaviours that we see in these other primates. We can find many visual examples in photographs and film clips that are readily available on the Internet: beachcombing along the margins of shallow lakes and on the coast for crabs and other foods, bathing and swimming (including completely submerged) and wading in shallow water, on all fours but also on hind legs when in deeper water in search of aquatic plants and animals, are behaviours that we can observe among a diverse range of primates. These include several species of macaques (genus *Macaca*), baboons (*Papio*), proboscis monkeys (*Nasalis*), swamp monkeys (*Allenopithecus*), capuchins

(*Cebus*), gorillas (*Gorilla*), chimpanzees and bonobos (*Pan*), orang-utans (*Pongo*), and even some species of lemur (*Hapalemur*). It seems quite likely that the early hominids got close to water and may have behaved in similar fashion to some of the primates that I have just listed.

So what can we conclude about our earliest hominid ancestors? They seemed to have taken an important step from fruit-eating canopy dwellers towards omnivorous facultative bipeds. They spent time on the ground and time on the trees. The new diet probably reflected the new world of these early hominids but it probably took the form of a gradual change in the proportion of food types consumed, maybe upping insects and other invertebrates and lowering fruit and other plant matter. The precise allocation of different foods to the dietary repertoire would have depended on where each group lived so we would expect differences between groups and times of the year. The food search patterns would have been similar to those of their fruit-eating ancestors, seeking patches of plenty in an otherwise barren environment. Once again, the new search patterns were a consequence of forest canopy life, transferred to a new setting.

I imagine Ramidus and Toumaï behaving a little like the wild boar of Doñana, hanging around the shallow waters and taking everything that became available. This opportunism may have started a new way of exploiting the environment: by taking a wider range of foods than before, they may have been able to survive without having to roam across great distances. But if drought came then they would have had to move across greater distances in search of other water bodies, just like elephants do today. In doing so they may have found ways of using their newly found tastes for variety by exploiting the bits of ground between one water body and the next. We shall look at this in detail in Chapter 3.

What of the Aquatic Ape? Have we found any evidence for an obligate aquatic primate just at the time that the hypothesis predicts? None whatsoever. We should not confuse the evidence of association with shallow waters, which are also used by many non-naked quadruped primates, with proof of a fully aquatic existence. The fossils of the early hominids reveal nothing that even hints at such a way of life. But this chapter has taught us another lesson: that the dichotomy set up by supporters of the Aquatic Ape Hypothesis between an aquatic or a savannah lifestyle, polarizing the debate around which of these best explains the evidence, is false. Our earliest ancestors were not living on the open savannah, but that does not mean that they were fully aquatic either.

3

At the Lake's Edge

4.2 TO 1.8 MILLION YEARS AGO

It was a hot September afternoon as we climbed to the top of the Rock of Gibraltar. With me were ecologists and ornithologists who had taken part in a conference on the evolution of bird migration. The climb was part of the post-conference field excursion. As we reached the highest, knife-edge ridge we could hear intense and incessant screeching. It was disturbing.

Once at the top we realized what was going on. A troop of Barbary macaques[1] had been approaching a second group and was being challenged. One group was arriving from the slopes below and their objective was clear: a pond with fresh water that was controlled by the second troop. These animals were so engrossed in their personal war that we were able to walk among them as if we were an inanimate part of the landscape. Females held the front line, infants screaming in fear as they clung on to their mothers' backs. The males appeared more reticent, seemingly trying to hold the females back. Time and again the rogue group would launch an attack, and noise levels rose terrifyingly as individuals exchanged blows and bites. Blood was shed. I turned to one of my colleagues and said, 'This is like watching our ancestors fighting over water in some remote African waterhole.' That is how it felt. There was something primal about the scene and it was just too close for comfort. After several hours the rogue gang retreated, dehydrated, with the infants unable to take any milk from their stressed-out mothers. I don't know how they fared in the days that followed.

September in Gibraltar is the height of the summer dry season. Typically, it does not rain at all between mid-June and October. The land is baked; there are few flowers and much of the ground vegetation is crisp and brown. Snails cluster on branches and have shut down all essential life-support systems as they try and bear the brunt of the drought. Hibernation in northern climates is a well-known phenomenon but the similar process of aestivation in southern lands, achieving a similar purpose, is not so familiar. It is hot, but it is water shortage that is the key limiting factor.

I was imagining an australopithecine[2] war when I experienced this confrontation but was there any factual basis to my thoughts or were they pure romanticism? Perhaps we can try to answer this question by establishing where and how the australopithecines lived. In Chapter 2 we established several key components of the habitat of our earliest ancestors. One of them was water, but we found no evidence to support the idea that they had been fully aquatic. How important was water to the australopithecines that followed them, and had their habitat and habits changed at all?

To resolve this issue I looked at habitat descriptions that had been published in association with the various species that have been recovered from the fossil record.[3] I separated the gracile australopithecines, usually placed in the genus *Australopithecus*, from the robust forms (genus *Paranthropus*) and the early forms ascribed by some authors to *Homo* (*habilis* and *rudolfensis*) and which I will refer to here as *Homo-Australopithecus*. Were there any significant differences in the habitats occupied by the different kinds of early hominids? Let's start by taking a look at *Australopithecus*.

Australopithecus includes such famous fossils as Lucy and several species[4] which occupied a wide geographical area of eastern, central, and southern Africa between 4.2 and 2.0 million years ago. For my analysis I have relied on two very useful summaries and additional individual descriptions.[5] Eleven sites give us useful information about the places in which *Australopithecus* lived and, as we can imagine, not everyone is in agreement. To avoid the onerous task of having to adjudicate on the merits of the various descriptions, I have simply kept to the approach that I outlined in Chapter 2: of looking at elements of the environment that appear to turn up constantly in the descriptions. Of the 11 sites, nine (81.8 per cent) had water present in some form. The descriptions that indicate the proximity of water include riverside woodland and forest, wet grasslands,[6] and lake margins.

All 11 sites give us clear signals that trees and open spaces were present. Rather than get involved in discussion of whether *Australopithecus* lived in forest, woodland, savannah, or grassland, or indeed of coming to the conclusion that they occupied habitat mosaics, what this approach does is tease out the main features of a typical *Australopithecus* habitat. Trees, like water, were essential components but so were open, treeless spaces. The proportion of each element probably varied significantly from place to place and between the different species but I do not think that we have enough information to fine-tune to this level of detail. What I do think we can conclude is that they lived neither in dense forest nor treeless plains.

Bushland, places dominated by shrubs, seems to have been frequented but not as much (six of the 11 sites) as places with trees and open spaces. Shrubs tend to produce dense thickets which may have been impenetrable and dangerous, places where predators could have lurked in ambush. The attraction of these places would have primarily been as sources of food, particularly fruit in season. But did *Australopithecus* eat fruit?

A battery of recent studies on dental microwear and stable isotopes[7] in early hominids has provided information about what they ate, with a few surprises along the way. For *Australopithecus* we now have data for four species: *A. afarensis, A. africanus, A. anamensis*, and *A. sediba.*[8] The first revelation was that, despite a morphology that had suggested that they consumed hard foods, dental microwear patterns showed the opposite: *Australopithecus* did not eat hard nuts, but they did not behave like modern leaf-eating primates either; tough leaves were not a part of their diet. One explanation that has been offered to explain this apparent conflict between a robust morphology and a relatively soft diet is that these hominids only went for hard foods in times of hardship. In other words they were fallback foods and they were taken at critical times only.[9]

Another important result was that the two East African species (*anamensis* and *afarensis*) showed little variation among individuals even though the specimens studied had been taken from areas that had been as far as 1,500 kilometres apart and were separated by as much as 1 million years.[10] So they seemed to have been quite consistent with regard to the types of food that they ate and this varied little across their geographical range and altered little over time in spite of climatic and vegetation changes. Then came *sediba*.

When Amanda Henry and colleagues studied this species[11] they were astonished because it was so different from the other species. One of the two *sediba* individuals that were studied had a microwear pattern in the teeth which indicated that it had eaten hard foods prior to death. It stood out from the other *Australopithecus* in this respect. A first in Amanda Henry's study was the identification of microscopic particles of silica deposited in plants, known as phytoliths,[12] in an early hominid. The recovered plant remains from the teeth provided direct evidence of the plants that *sediba* had been consuming. *Sediba* had been eating fruit, leaves, bark or wood (which contain protein and soluble sugars), shade- and water-loving C_3 grasses and sedges. The sample was small but clearly indicated a varied plant diet. The absence of C_4 phytoliths, which were common in the sediment from which *sediba* had been recovered, showed that *sediba* had avoided such plants just like modern savannah chimpanzees do. This result contrasted with the bigger picture for *Australopithecus*, which was of variable consumption of C_4 grasses but at higher proportions than savannah chimpanzees or *sediba*. The regular consumption of C_4 foods has been regarded as a fundamental trait, along with bipedalism, that permitted these early hominids to pioneer increasingly open and seasonal environments.[13] *Sediba* was different.

So, the answer to the question of whether *Australopithecus* ate fruit, and hence entered bushland to obtain them, would depend on which species we were considering. *Sediba* clearly did, and the other species would probably have done so as well but were not as reliant on fruit as *sediba* would have been. So we would expect

Australopithecus to have entered areas covered in bushes at times when fruit where available but it would not have been a feature of their regular haunts. The eating of sedges by *sediba* at least would indicate that it lived close to water and took aquatic plants. The other *Australopithecus* species also lived close to water and may well have eaten similar plants even if they were not as committed to them as *sediba*.

We are beginning to form an image of *Australopithecus* in its world but we need to answer one more question, did they walk upright? The answer to that question seems to be that they were indeed striding bipeds, much more committed to this method of locomotion than their predecessor *Ardipithecus*.[14] The fossil tracks at the famous site of Laetoli give clear confirmation of bipedal walking even if the makers of the footprints, generally attributed to *Australopithecus afarensis* (Lucy), may not be known with complete certainty.[15] But was *Australopithecus* also capable of moving freely in the trees, as *Ardipithecus* had done? This has been debated for a considerable time and the anatomy of the foot raised doubts about how capable *Australopithecus* would have been as a tree climber. *Sediba*, once again, set the cat among the pigeons. Studies of the ankle, foot, wrist, and hand of this species suggested that it was indeed bipedal but also partly arboreal.[16] It seems that not all *Australopithecus* species necessarily behaved in the same manner.

So what picture can we paint of *Australopithecus* between 4.2 and 2.0 million years ago? I see a group of sociable species that walk upright on the ground, usually close to trees which they rely on for cover, safety, and sometimes food. They probably sleep in the trees. *Australopithecus* are rarely far from water and they probably wade in search of sedges and other aquatic plants. It is unlikely that they ever venture right into the open but they are equally wary of dense vegetation, avoiding forest and cautiously venturing into bushy thickets when searching for ripe fruit. The choice of place in which to live is varied within the limits that I have set out—places with trees, open spaces, and water in varying amounts—as is their diet which may include C_4 grasses, fruit, and harder foods, the latter at least in times of shortage. I will return to *Australopithecus* at the end of this chapter but first we need to have a look at the robust australopithecines and early *Homo*.

The robust australopithecines have been traditionally placed in a separate genus, *Paranthropus*. They appear later than *Australopithecus* in the fossil record and survive later too, overlapping for a long time with *Homo*. We can place the time frame between 2.8 and 1.4 million years ago.[17] They had a wide eastern African geographical distribution, from Ethiopia right down to South Africa. Using the same sources of information as for *Australopithecus*, I was able to find 21 sites that provided useful information on *Paranthropus* habitat. There was little variation between these robust australopithecines and their gracile cousins in proportion of habitat elements, with one exception. Water was again a key factor in *Paranthropus* habitat (95.2 per cent of all sites had indicators of presence of water). Trees and open spaces were also important and bushland much less so, as in the case of *Australopithecus*.[18] One feature stood out and may be significant. It was the number of sites recording open habitat and no trees. There were five such sites, which may seem few but represent almost a quarter of all the sites and contrasted with the *Australopithecus* sites that always had trees present. Does this mean that these later australopithecines were moving away from the trees? It seems so and bears out Reed's[19] conclusions that *Australopithecus* species existed in fairly wooded, well-watered regions while *Paranthropus* species lived in similar environs and also in more open regions, but always in habitats that included wetlands. Perhaps in habitat terms, we are seeing a parallel between *Paranthropus* and baboons.

These robust australopithecines may represent a lineage that ventured further into the open plains than any other before them. The southern African *robustus* seems to have had a wide and varied diet with clear evidence that it was capable of eating hard foods. It is possible that they only went for these as fallback foods when softer ones were scarce, in the same manner as some modern monkeys do.[20] In contrast the East African *boisei* stands out among all early hominids in the high intake of C_4 foods, in the region of 75–80 per cent of its diet, very similar to the grass-eating warthogs, hippos, and zebras. Such a large intake would certainly seem to fit a hominid that was venturing more into open grassland environments.

Paranthropus, literally meaning 'beside man', was a parallel lineage to humans which went extinct. It was not our ancestor. It illustrates the degree of evolutionary experimentation that was happening over a period of millions of years as forests shrank and grasslands expanded. Like *Australopithecus* its movements appear to have been tied to the proximity of water, but *Paranthropus* seems to have started on a path that moved it further away from trees, without abandoning them altogether. We now need to look at a third group which, like *Paranthropus*, were descendants of an ancestral lineage of *Australopithecus*. This third group is *Homo-Australopithecus*.

Homo-Australopithecus contains two species: *habilis* and *rudolfensis*. These hominids resembled the australopithecines in size, including brain volume, and some authors have considered them to be so similar that they have placed them in *Australopithecus*.[21]

Since they are generally regarded to be in our direct ancestry, we shall take a look at them separately from the other two groups. Their time frame is between 2.33 and 1.4 million years ago, overlapping almost completely with *Paranthropus* but not with *Australopithecus*, except the late surviving *sediba*. Their geographical distribution matches *Paranthropus*, from Ethiopia down to South Africa.

Nine sites provided information on where early *habilis/rudolfensis* lived and they appear to differ little from *Australopithecus* or *Paranthropus*. They too depended on water, trees, and open spaces.[22] One of the nine sites did not have trees and another did not have water, suggesting very tentatively that, like *Paranthropus*, *habilis/rudolfensis* may have been venturing further away from cover and water sources. It may reflect greater mobility among these bipedal hominids, which would have allowed them to stray further and further from these key places in the knowledge that they had the capability of returning to them with relative ease. They may have still returned to the safety of trees at night but were prepared to venture into the open and away from water in search of food. Some died in these open, waterless places, as the discovery of their remains in these contexts indicates. But overall, trees and water would never have been very far away. The diet of *habilis/rudolfensis* seems to have been unspecialized and was not geared towards either hard or tough foods. Instead it seems that they had a fairly generalized diet that included C_3 and C_4 foods in similar proportions.

When we put all this together we can begin to assemble a possible chain of events which is the history of these early hominids. From the appearance of the first australopithecines to around 2.8 million years ago, when we find the first *Paranthropus*, the picture is not too different from the earlier one described in Chapter 2. It is one of hominids walking on two feet in areas with trees, open spaces, and water close by. They ate a range of plant foods but largely avoided hard nuts and similar items. Then, something seems to change after 2.8 million years ago.

The African climate took a shift towards becoming drier and more variable in the time interval between 2.9 and 2.4 million years ago, the first of three such shifts which were related to the start and strengthening of the influence of glacial cycles in the tropics.[23] One consequence seems to have been the opening-up of areas of woody vegetation which became dominated by wooded grassland.[24] The open grasslands, which we associate with present-day Africa, came much later[25] which means that we cannot implicate them in these early steps in our evolution. The changes that climate generated around the 2.8-million-years-ago mark were in the direction of expansion of open spaces within a framework of woodland vegetation. In other words a patchwork or trees and open spaces was the outcome, exactly what the australopithecines had been exploiting for close to 1.5 million years. It would have generated a population boom which seems to have been accompanied by the start of the new *Paranthropus* line at 2.8 million years ago and *habilis/rudolfensis* at 2.33 million years ago. The different australopithecines were splitting the wooded grassland cake between them.

One additional consequence of these climate changes would have been a reduction of the area covered by wetlands which probably became more seasonal in character as well. We have already seen just how important water was to the australopithecines and new seasonal shortages may have given rise to stress and mortality which would, in turn, have intensified the pressure in favour of adaptations that improved survival under these new conditions. Improvements, for example in the hind limbs, that allowed hominids to move quickly and over greater distances across the landscape, would have been seized upon rapidly by natural selection. This was, in my opinion, *the* trigger to our evolution: adaptations which promoted mobility were spurred on, above all else, by the need for swift and efficient movement between ever-shrinking sources of water. As the pressure increased, so compromises had to be made at the expense of life in the trees, and the anatomy of the terrestrial biped was fine-tuned. The start was made in some tropical African wooded grassland at a point prior to 2.8 million years ago and it took off after this time. Once started, there was to be no return to the trees. Such unconditional commitment to life on the ground was not a feature of *Australopithecus* but it seems that *Paranthropus* and *habilis/rudolfensis* may have been edging closer to it.

I want to end this chapter with three aspects that were important during this initial period of arboreal emancipation: rocky places, stone tools, and meat. Let us start with rocky places. I looked for evidence of rocky habitats—caves, cliffs, screes—in the habitat descriptions of the australopithecines that I have used to build a picture in this chapter. I found evidence of such habitats in 18.2 per cent of the *Australopithecus* sites, 23.8 per cent of the *Paranthropus* sites, and 22.2 per cent of the *habilis/rudolfensis* sites. There was no evidence of rocky places in *Sahelanthropus* or *Ardipithecus* sites so it seems as if we have a new feature here, one that came with the australopithecines. The number of sites reflecting rocky habitats is small in comparison to other habitat elements—trees, open spaces, water, even bushes. Rocky places may not have been essential to the australopithecines, which seem to have retained an ancestral desire to seek cover among the trees, but it is there and we should account for it.

A recent study has proposed that places in eastern and southern Africa that were subject to repeated tectonic or volcanic activity may have been favoured by early hominids such as *Australopithecus* because they offered situations in which heterogeneous or mosaic habitats were frequently found as a result of the abrupt changes in relief caused by geological processes.[26] Of course, when they talk of heterogeneous or mosaic habitats or landscapes they really mean places in which a variety of habitat elements were present in close proximity and in varying proportions. We should be quite happy by now that such places were what the australopithecines actively sought out. They were not habitat mosaics in the eyes of the australopithecines but instead were what they perceived as their habitat. These places would have provided ample water, as streams and rivers cut into the newly exposed rocks. As these rivers flowed downstream, different geological settings would have generated lakes, swamps, and other wetlands. They would have also left behind cliffs of harder, uneroded rock. It is this latter aspect that interests me most because the rest of the key habitat elements could have been found in other areas which were not so geologically active. But venturing into these attractive areas of tectonic and volcanic activity brought australopithecines close to significant areas of exposed rock. In these areas there would have been caves.

We have established that the climate started to get drier after 2.8 million years ago. Water sources were more distant from each other than before and I have proposed that it was this increased patchiness in water sources that promoted early hominid mobility. But cover in which to sleep in relative safety at night became increasingly difficult to find, too, as trees gave way to open spaces. Cliffs and other rocky places would have provided alternative locations of relative safety from predators and would have also given the early hominids new options. Imagine a rich wetland, full of food and with plentiful water but in a situation that was far away from clusters of trees where the early hominids felt safe. Going to such a wetland would have meant taking big risks, moving over open areas in which they would have been exposed to dangerous predators. Then having to go back and forth between trees and water every day would have meant that a lot of time and energy was wasted. Imagine further that a line of cliffs overlooked the wetland. That would have solved the problem for an adaptable hominid that was able to take the unprecedented step of choosing to sleep in a place that had no trees. Cliffs, caves, and other rocky places permitted the australopithecines to exploit new territories that would have otherwise been out of reach. Rocky places were woodland substitutes.

Should it surprise us that the earliest known stone tools made by early hominids should date to 2.6 million years ago, precisely at the time of the first big drying event in Africa?[27] The australopithecines must have been using tools long before this time, just as chimpanzees do,[28] made from wood which they had readily accessible, and they may have used stones, too, when these were available judging from recently discovered marks left on butchered bones in Ethiopia which are estimated to date in the region of 3.42–3.24 million years ago.[29] It is possible that, prior to 2.6 million years ago, they simply picked up isolated stones or they may have, instead, deliberately visited known quarries where they took particularly attractive and suitable pieces. We just do not know to what degree they shaped them into tools either. But by

2.6 million years ago they were certainly turning rocks into tools and visiting places where suitable rocks were frequent and this behaviour may have made them familiar with particular rock sites that they then started to use for other purposes, such as shelter.

The use of tools by the early hominids has been linked to a change of diet to include meat and other mammal products, like bone marrow. Tools, it is said, opened up the niche of the early hominids by allowing them access to foods that they could not have been able to process with their teeth. Some authors have gone as far as to make meat the central ingredient that made us human.[30] I am not convinced and will examine this contention in some detail in Chapter 4. We have seen here that the australopithecines ate a wide range of plants. *Sediba* has shown us how wide that choice really was. We have to assume, in the absence of direct evidence, that this extensive plant diet was complemented by related animal foods if present-day monkeys and apes living in similar situations are anything to go by. It does not take a huge mental leap to imagine an australopithecine picking up molluscs while wading in shallow water for aquatic plants, picking up a stick and digging into a termite mound, or stalking an unsuspecting lizard basking on a rock in the morning sun. That they would have stumbled across carcasses when venturing into open spaces has to be a given. It does not take much to imagine that the primate brain brought down from the forest canopy incited its owners to explore the content of these unusual finds and to figure out how to get some food from them. Did this ability extend to chasing live animals or to chasing dangerous predators away from their prize? We cannot be sure. I remain sceptical but I do accept that it would have been the next step, and that it was imminent.

4

The First Humans

1.8 MILLION TO 800 THOUSAND YEARS AGO

On 29 August 1985, the journal *Nature* reported the discovery of the skeleton of a boy who had died at the age of 12 around 1.6 million years ago. It was the most complete skeleton ever found of *Homo erectus.*[1] The tall youth, with a stature estimated at 1.68 metres, became known worldwide as the Nariokotome Boy, after the name of the site in which he had been discovered. The site was on the south bank of the Nariokotome River in Kenya, close to where it drained into Lake Turkana on its western shore.

Homo sapiens erectus (as I am regarding them as a subspecies)—the first humans—came onto the scene around 1.7 million years ago, almost simultaneously in East Africa and south-east Asia, making its area of origin hard to determine.[2] Some scientists prefer to separate the two populations, leaving the name *erectus* for the Asian population and classifying the African population as *ergaster.* In my earlier book, *The Humans Who Went Extinct,* I did not draw the distinction. I will stick to that preference in this book. In Africa, humans would have lived in similar geographical areas as *Paranthropus* and *Australopithecus habilis/rudolfensis* for around 300 thousand years but only humans survived after 1.4 million years ago.

The fauna recovered from the Nariokotome site, which included fish, crocodiles, turtles, elephants, and hippos, clearly indicated the presence of water where the Nariokotome Boy died, and had presumably lived. In the database for the australopithecines which was used in Chapter 3, seven African, including Nariokotome, sites were included from which I could tease out information on the habitat of the first humans. To my surprise no important habitat differences between these humans and their predecessors and contemporaries could be found.[3] In fact, trees were always present in human sites from this period which put paid to any suggestion that this hominid had finally taken to the open, treeless plains of Africa. So, if humans lived in similar places to the australopithecines, what made them different enough to be able to persist among a suite of similar species without suffering the effects of competition? And if they did not have the edge over potential competitors—they seem to have lived alongside each other for close to half-a-million years—why were humans the only ones to survive after 1.4 million years ago? The answer, I will argue, is water in a world that was drying up.

The period between 1.8 and 1.6 million years ago saw the second of three major climatic downturns which were superimposed onto an overall climatic trend of cooling, drying, and increasing climate variability.[4] The African climate became even drier and more variable than before and there were significant changes in the fauna. *Homo sapiens* appeared at this juncture. The effect of the climate changes on the land would have been significant and would have been expressed visibly by a landscape in which open spaces were becoming increasingly large and common at the expense of trees. But areas which until then had remained covered by rainforest would also have given way now to open landscapes with trees. So what we would have observed would have been a loss of rainforest, perhaps little change in areas with trees and open spaces, and an increase of areas of open spaces in which trees were scarcer than before. I suggest that the areas with trees and open spaces may have remained similar because the losses in favour of open areas in the drier regions may have been offset by new areas which were once rainforest. Water sources would have become patchier still and seasonal effects would have become even more pronounced than before. Dry and wet seasons would have alternated.

This is the scenario in which we first find *Homo sapiens*. In *The Humans Who Went Extinct* I put forward the idea that the thrust of evolution was happening in marginal areas where populations were most stressed. In core areas, populations stuck to what they knew best and it worked—provided conditions did not change. Life was tougher on the margins and death and extinction were frequent. But when conditions changed in the direction of those that had been experienced on the margins, then these peripheral populations got their chance as the conditions in which they had managed to scrape a living spread. Humans would therefore have started off within marginal populations of *Australopithecus habilis*.

How do the newly arrived humans on the scene compare physically with *habilis*? This is an important question which might tell us something about their respective ways of life. A recent analysis of the skeleton of the Nariokotome Boy, for example, has revealed that it was not as tall as previously estimated.[5] It now seems that the Nariokotome Boy had a growth spurt that did not reach the speeds of present-day humans and it ended sooner too, around 12.3 years of age. The earlier models that had been used to calculate growth rate, and from that its stature, were wrong. The Nariokotome boy was 163 centimetres (5'4") and not 185 centimetres (6'1") as had been thought for a long time. This new estimate fits in well with other estimates of the stature of humans from this period which are based on the lengths of fossil femurs and which fall in the 160-centimetre (5'3", female) to 180-centimetre (5'11", male) range. These estimates would still make humans considerably taller than *habilis* whose females stood at 125 centimetres (4'1") and males at 157 centimetres (5'2"). *Australopithecus* and *Paranthropus* were even shorter. McHenry, who studied stature in early hominids,[6] concluded that it was 'not true that humans have been getting progressively taller throughout their evolutionary history. Some individuals were as tall as modern [meaning recent] humans 3 mya [million years ago], by 2 mya one individual stood about 173 cm, and by 1.7 mya a stature of 180 + cm was not uncommon.'

Humans were also significantly heavier than *habilis.*[7] The body weight increase from a male *habilis* to a male human from this period may have been of the order of at least 20 per cent but it is in the females that we observe the sharpest weight increase, which was of the order of at least 50 per cent. Prehuman males increased slightly in size through time within the range 40–52 kilograms (88–114 pounds) while females did not and stayed in the range 29–34 kilograms (64–75 pounds) throughout. So humans marked a step in terms of body-size increase and it was most pronounced among the females of the species. Consequently, the sexual dimorphism in body size, so evident in the australopithecines, was significantly reduced in humans.

Large size has important implications in ecology. I am interested here in the effect of size on the movement of animals. It has a direct bearing on net energy expended in travel, net transport costs[8] decreasing with increasing size. It is interesting that body size has the opposite effect on climbing animals,[9] which may have restricted body size in hominids until they committed to walking on flat surfaces. Larger animals also reach higher maximum velocities than smaller ones when running; they move at faster optimal speed (the normal speed of movement) and they therefore cover greater distances in the same amount of time. This means that humans would have done better than *habilis* in terms of energy efficiency as they walked across their territory and they could have moved faster if required to do so. They could also have had a larger home range.

Hind limb proportions also improved the efficiency with which humans moved. Increase in body size meant that the total cost of locomotion also increased, even if the net cost (energy expended per kilogram) decreased. When the daily cost of movement was calculated for a human from this period with its elongated hind limbs, travel costs were 50 per cent lower than for a hypothetical human with the hind limb proportions of an australopithecine.[10] The elongated limbs of humans mitigated to some degree the increased cost of locomotion due to its larger size, making walking more economical than in the australopithecines.[11] Longer hind limbs in humans have also been found to reduce running costs.[12] The lengthening of the hind limbs in humans improved their efficiency not only as walkers but also as runners. *Australopithecus habilis*, of similar size to the australopithecines and hence with a more economical, though less efficient, walk than humans, would have had its own advantage over other *Australopithecus*, by virtue of having longer hind limbs. We can see how, after a long period when hind limb proportions remained quite constant among the australopithecines, the increase that we first pick up in *habilis* and which really took off in *Homo sapiens*, especially in the earliest *erectus* humans, would have dramatically augmented the daily distances that could be covered. We have seen that the habitat requirements of early humans in tropical Africa were essentially those of their predecessors. What really changed in humans was the ability to cover much more ground in a day and to roam across larger home ranges than any of their predecessors. Such a change was the result of an increase in body size and the length of the hind limbs. Once committed to this new road, any further increases in these features would have been favoured by natural selection. Standing tall in the East African landscape 1.8 million years ago were the first humans. From an ancestral, fruit-eating, forest-canopy primate millions of years of evolution assisted by climate change had finally produced this improbable primate.

Why were humans so improbable? The world of humans and their predecessors had been dominated for millions of years by a wide range of mammalian herbivores[13] and a guild of carnivores.[14] The australopithecines were not alone among the primates in this world either, terrestrial baboons being noteworthy neighbours[15] alongside pigs which were also omnivorous.[16] All these species were perfectly capable of consuming meat, even the pigs,[17] but they would have done so opportunistically or at a small scale relative to the dominant and specialized carnivores. They had reached a level of improbability, though not uniqueness, but a minor one in comparison to humans who seem to have barged into the carnivore guild in their own right. That, for a primate, was certainly an unexpected first.

What pushed the ancestors of humans in the direction of increased mobility and how did that generate the world's first and only primate carnivore? It had to be water. As water sources became scarcer, more thinly scattered on the ground, and increasingly seasonal, hominids would have been faced with three choices. The first was to stay put close to water and ride out the bad times while eating what you could find around you—a good option if you were not too bothered about what you ate but one in which you might have to depend on low-quality fallback foods for a while. It carried the added risk that the water source might dry out altogether. Second, you could stay close to water but you could range more widely than in the first option, provided you were able to return to the water source. This option would certainly permit you to exploit a wider range of resources but you would still run the risk of the water source disappearing in extreme drought. The third option would be to become highly mobile and move between two or more water sources during the annual cycle. Such a strategy would seem the least risky and would offer many more dietary options. It would also put the users under intense pressure from natural selection, favouring any genetic novelty that promoted improved mobility.

It is time to introduce a new concept, one which has been discussed occasionally by ecologists and which is vitally important to our story—it is usually known as the Baldwin Effect.[18] It happens when an animal learns a new behaviour, either by itself or from others, and this behaviour eventually becomes incorporated into the genome of future generations if it confers advantages. Such novel behaviours can place individuals in new environmental settings which may, in turn, expose them to new selective pressures. What we are looking at here are ways in which the behaviour of an animal could affect the course of evolution. I do not mean, of course, Lamarckism, the discarded view exemplified by the animal that stretched its neck each time further to reach the leaves on the highest branches and produced a giraffe after several generations. There are ways in which animals can position themselves in particular locations or even modify their environment. Patrick Bateson,[19] the Cambridge animal behaviourist, usefully summarized these possible ways into four. The first was when the animal made an active choice of where to live; the second was by changing the physical or social conditions, through their behaviour, in which their offspring would live; the third was by changing their behaviour when conditions changed, thus avoiding death; and the fourth, especially relevant to us, was when animals exposed themselves to new conditions which could open up possibilities for evolutionary changes that would not otherwise have taken place.

Let us make what we are saying here relevant to humans. For example, let us assume that an ancestral population of this species, deliberately or because they had no choice, moved into empty territories that were not exploited because water sources were very distant from each other. If that population had genetic variability which had the potential to have larger individuals with longer limbs, then natural selection would favour the larger, longer-limbed individuals. In time a larger, longer-limbed population would be exploiting this territory and would have become more efficient at doing so. We can contrast this situation with a second population that stayed put, had the same genetic variability, but was not under pressure to move in that direction.

By placing itself in the new conditions, the first population was unconsciously making itself available to evolutionary change. More often than not, such attempts at exploitation of marginal areas probably end in disaster and those are the myriad stories that we cannot tell. Occasionally, there is success and we have a tale. Our narrative is one of those few.

I have argued for water as the principal driver of the evolutionary changes that characterized the first humans because it was a constant requirement for their daily lives. *Homo sapiens* was an evolutionary response to the scattered distribution of water in space and time. Food and shelter would have been secondary drivers for the simple reason that the early hominids were not committed to particular food types and had, as we have seen, found shelter alternatives to trees in the world of rocks. What was not open to alternatives was water, which is why it had to have been the main limiting factor and the principal driver of our evolution. Improved terrestrial mobility was a response, first and foremost, to the need to quickly locate water sources in a drying world.

Where are we so far? We have a large, active, and highly mobile terrestrial biped that occupied similar habitats to its predecessors. It had a larger home range which must have exposed it to a wider range of situations than its predecessors who lived within the confines of a smaller territory. The movement of the first humans may not have been limited to a day range. For the first time in the history of this primate's lineage, movement probably involved seasonal changes of location within an annual home range. Such movement would have further increased the probability of encountering a diversity of food sources. For a hominid that had already developed a taste for meat, marrow, and fat, its mobility would have increased the chances of encountering carcasses which would have been ideal alternative sources of nutrients. Mobility would have facilitated quick arrival at fresh carcasses, by observing the flight behaviour of vultures, and their large size and group behaviour may have allowed them to compete with the strong predators and scavengers that lived in these places.

We do not know when humans started to hunt as well as scavenge. Perhaps their predecessors already hunted but, if so, it may have been on a sporadic or opportunistic basis. We cannot tell simply from cut bones. I suspect that hunting herbivorous mammals became a regular feature of the repertoire of human behaviour at a very early stage. Hunting and scavenging for meat and related products were novel behaviours that would have been open to the Baldwin Effect. Hair loss and sweating were probably products of hunting and scavenging.

The areas that humans frequented were rife with powerful and dangerous predators. These predators would have been most active at dawn, dusk, and at night, just like those that live in tropical Africa and India today.[20] One tactic that smaller predators use to avoid the more dangerous, larger predators is to shift the time when they are active in order to avoid encounters.[21] Curiously, it is the smaller predators that hunt mainly by day[22] which could be taken to mean that the dominant predators have taken up the prime time slots, which would be those that provide most cover and avoid the intense heat of the middle of the day. Cheetahs survive by living in what have been described as competition refuges—areas of low prey density that are unattractive to the larger predators. Combined with different feeding times, cheetahs, whose densities are directly controlled by the larger predators, are able to persist in heterogeneous environments that offer such competition refuges.[23] If humans wanted to avoid the most dangerous predators, they would have been well advised to limit their activities to those times when the predators were resting or sleeping, that is, during the heat of the day, and to those places where they were least likely to encounter them.

Such a strategy would have been useful simply as a means of escaping being eaten and that may well have been how the behaviour started. Once behaving in this manner, humans would have found themselves in a position in which they had a good chance of taking food from carcasses, when the main competitors would have been vultures. After that, it would have been simply a matter of time before they started actively hunting. Predator and competition avoidance drove humans towards a life in the middle of the day. But such behaviour put this bipedal terrestrial primate in a position in which it was at risk of overheating. The selective pressure to deal with this would have been huge and any variability in the human population that alleviated the problem would have been seized upon. Becoming naked and adopting a cooling system that involved the production of sweat were the answer.[24]

The defenders of the Aquatic Ape Hypothesis argue that the system is inefficient because it means that a hominid in a hot environment needed to conserve water and would not have developed a sweat-production cooling mechanism. This is a fallacy that seems to assume that Nature is perfect. It is not. We have seen how much of what we observe is a by-product of other developments; organisms do the best with what they inherit, and compromises are frequent. A naked, sweaty hominid may not have been the best solution in an ideal world but it was the solution for the state of affairs of the world of 1.7 million years ago. Many other animals, including some primates, lived successfully in that world and sorted their overheating problems in other ways. Comparing the different ways and trying to find which was best is futile. They were all good because they allowed the different species to survive and reproduce. What happened to our bare and perspiring hominid was that, in order to be able to make a living in its hot world, it tied itself down to having to drink water frequently, and this need would have put further pressure on being highly mobile and able to move quickly from one water source to another. Natural selection drove a ratcheted linked response in which preventing overheating and preventing dehydration fed off each other.

I will return to nakedness in Chapter 5 in the context of the spread of the first humans away from tropical areas. In this chapter I have attempted to build a picture of how and why humans were a viable proposition 1.7 million years ago. Mobility was at the heart of this new primate solution to a drying world. It should not surprise us then that, after close to a million years of observing tools—known as Oldowan—that appear to have been of a kind that were readily made and discarded *in situ*, precisely at this juncture we should find a new technology—the Acheulian—which seems to have been the hallmark of mobility.[25] From the very origins of this new technology, tools were either made with thick, pointed tips or were long with durable cutting edges. The first type seem to have been used as picks for tasks that needed bulk and weight while the latter provided stable cutting edges. These tools were probably transported between sites by humans and they would have had greater scope when roaming across their large home ranges. Carrying tools with the freed hands of a biped would have provided an instant match between newly discovered food and the ability to process it immediately. Prior to this new technology, early hominids would have often found themselves in situations with food but without tools that gave them access to it. Sometimes they may have been able to carry food back to quarries or stashes of tools with all its risks, such as meeting powerful competitors along the way. The ability to carry the tools with them was an important step in the new world of mobility.

Manufacturing tools of this type required advanced motor skills and cognition. The first humans clearly had brains that could deal with these new problems but a larger brain also brought with it new problems. In this chapter I have kept the focus on humans in their tropical African home at the very beginning of their emergence. In Chapter 5 I will widen the geographical and time frames to see what happened next. Part of that story will focus on the developing brain of this improbable primate that we can now identify as distinctly human. Not surprisingly, water remains central to that story.

5

Middle Earth: The Home of the First Humans

1.8 MILLION TO 800 THOUSAND YEARS AGO

In Chapter 4 we established that the earliest recorded African humans lived around 1.7 million years ago. In this chapter we look for their home in that period. Recent revisions place the actual date of these first Africans between 1.7 and 1.65 million years ago.[1] This means that the earliest Africans recognizably belonging to *Homo* are, on current evidence, around 100 thousand years later than the earliest Asian specimen, which is dated to around 1.8 million years ago.[2] These results, which could rapidly change with the discovery of new specimens, have put the location of the centre of origin of *Homo sapiens* in question:[3] were they in Africa as traditionally held, or were they in south-east Asia instead, or indeed somewhere in between?[4]

Dmanisi in Georgia is a priori an odd locality in which to expect early hominids. Well north of the tropical belt, at 41°N, and at an altitude above 1,000 metres, we would hardly have expected to find hominid fossils here, and certainly not of an age comparable to the earliest African humans. Yet, that is exactly what we have in Dmanisi—the remarkably well-preserved crania and associated remains of hominids that lived there around 1.7 million years ago.[5] Were they humans? We do not know. I am not even sure whether they belong to the *Homo-Australopithecus* group with *habilis* and *rudolfensis* or the earlier *Australopithecus* (but see Preface xviii). Taxonomic issues aside, there is an important point that we need to take home at this stage: by 1.7 million years ago early hominids were widespread across a vast area of northern, eastern, and southern Africa, central and south-east Asia, and we presume they also occupied the vast area in between. I have called this the mid-latitude belt[6] and Dennell and Roebroeks[7] called it Savannahstan. Here I will refer to it as Middle Earth—the home of the first humans and their contemporaries.

How did humans get to occupy Middle Earth and where did they come from? No clear answer has emerged so far to the second question. Tropical East Africa has traditionally been cited as the place of origin and the subsequent geographical expansion of humans has been labelled the 'Out-of-Africa 1' dispersal. My own hunch, given that the human adaptation was a response to a drying world, is that the area of origin was the core of Middle Earth and not the wet tropics of south-east Asia. That would place the area of origin of *erectus* somewhere in present-day north-east Africa, the Arabian Peninsula, the Middle East, or the low-lying coastal areas of the Persian Gulf.

As to the first question—how humans occupied Middle Earth—the first thing that we have to have very clear is that it had nothing to do with having elongated hind limbs, being highly mobile, or having large home ranges. I want to clear this matter up first because it has often been cited as the reason for the expansion of the first humans out of Africa. The anatomical and behavioural changes that came with the first humans, and which I described in Chapter 4, permitted their survival and reproduction. These were not features that directly influenced large-scale geographical expansion. They did so indirectly by facilitating the growth of the human population which in turn forced a demographic expansion as home territories were saturated and surplus individuals had to move elsewhere. The first geographical expansion of humans, like that of all other hominids and indeed all other animals, happened over a series of generations and did not involve the migration of long-legged individuals. Long legs were a factor promoting population growth, just as a powerful set of teeth might have been for a hyena.

The Dmanisi hominids were supposedly short and have been compared to *habilis*, apparently lacking the traits that characterized long hind limbs of humans.[8] Given what I have already said, my view is that if so, this is an observation that does not really have a bearing on the geographical expansion of humans or any other hominid, Dmanisi ones included, across Middle Earth. In fact, it now seems that the Dmanisi hominids were not short at all and a comparison with recent hunter-gatherers shows that they were very similar.[9] The real difference in stature only lies with some of the largest female early human specimens from East Africa. Those aside, there is a significant overlap in stature among the hominids of 1.7 million years ago, making the separation of those at the lower end of the stature range a difficult task. This problem has a bearing on the study of the brain of the early hominids.

The first humans had large brains but they showed considerable variation. The specimens that have been studied reveal brain volumes that range between 691 and 1,030 cubic centimetres.[10] This means that at the lower end the brains of the first humans overlapped with the larger *habilis* brains and at the upper end they overlapped with the smaller brains of modern people today.[11] Factoring body size into the comparisons, Kappelman came to the conclusion that relative brain size increased sharply in the first humans and then remained fairly constant. What did change after our appearance was body size, and brain size increased correspondingly. The only significant change to this trend seems to have occurred much later, around 100 thousand years ago, when there was a decrease in body mass with a correspondingly sharp increase in relative brain size. I will discuss these later developments further on in this book. What is important here is that *habilis* seems to have started a process of brain enlargement which continued in the first humans simply because they became larger. This means that humans had larger brains than *habilis* when looked at in absolute terms, though there was some overlap between sizes of the two.

Large brains require a lot of feeding and the enlarged brain of the first humans took a great deal more food than the brains of its predecessors, which it did at the expense of other organs of the body, particularly the gut which was reduced in size.[12] But the brain is also a sensitive organ that not only needs physical protection inside a hard skull but which must also be kept within a narrow working temperature range, constant and warm to maintain optimal activity of the neurons, if it is to function properly.[13] A core body temperature of 37 °C is optimal for all mammals and birds so it has deep evolutionary importance. To keep a lower core temperature would mean constant sweating and a higher temperature would be too close to the point at which proteins are denatured, risking vital life processes: so 37 °C seems to have been set as the ideal core body temperature.

The early hominids that lived in the hot conditions of tropical Africa would have faced a battle to keep cool and maintain a stable body temperature. Being larger than their predecessors meant they would have had a lower surface-area to volume ratio, proportionately less area of skin for their body mass, so problems of heat loss would have increased. The human solution would have been sweating, with the result that their activities would have been tied to sources of water. It has been suggested that humans needed large brains in order to be able to retain information on the location of water sources, particularly in sporadically occurring periods of extreme drought.[14] Certainly present-day Bushmen are unsurpassed among predators or prey in terms of the amount of information that they store and retrieve about the environment which they use during the course of a year—typically an area of 10 square kilometres. They retain an intimate knowledge of the terrain, especially of the water sources that they have previously used.

In Chapter 4 we saw how the appearance of the Acheulian technology seems to have coincided with the first humans. This new technology did not supplant the earlier way in which stone tools were made[15] and both have recently been found in the same sites[16] which suggests that they may have even been used by the same people but for different purposes. In Chapter 4 I also linked the Acheulian with mobility; these were tools that could be carried from place to place. What is even more interesting is that Acheulian tools almost always occur in association with fresh water.[17] We can now put together the early human package, or better still the package that defined the origins of humanity and the making of the improbable primate.

The package had a legacy that came from a rainforest fruit-eater. That legacy provided a brain capable of getting its bearer around the forest, remembering where and when to find trees with ripe fruit. The legacy also provided a grasping hand and a predisposition to stand on the hind legs when moving on horizontal branches. We find these three features in enhanced, even exaggerated, form in the first humans. The association with water came later and probably became indispensable as foods with high water content, such as fruit, were replaced by others that were drier, such as nuts. The change, which may have been initiated by *Sahelanthropus* or *Ardipithecus* or one of their generation, meant that drinking fresh water became part of the daily activities of the human prototype.

What types of places did early human populations inhabit? Let us explore the evidence for the early populations and for the pioneers that arrived in various parts of Middle Earth. In Java in south-east Asia, the human presence is as old, if not older, than in East Africa and the colonization of the island would have taken place via land bridges that are now submerged.[18] So what were humans doing on Java?

Humans occupied the site of Trinil in Java around 1.5 million years ago. Over 400 thousand fossils of aquatic and terrestrial vertebrates as well as aquatic invertebrates were excavated from this site during the last decade of the 19th century and the first decade of the 20th century and have been used to reconstruct the environment in which humans lived at the time.[19] The impressive list included fish, such as perch, catfish, sharks, giant freshwater stingrays, and sawfish, which indicate the presence of lakes, ponds, rivers, brooks, shallow waters, swamps, estuaries, and coastal waters. This conclusion was backed up by a large diversity of molluscs typical of such aquatic environments. The mammals from Trinil also revealed a close connection with aquatic habitats and the flooded meadows and forests that would have been close by. They included strong swimmers like the tiger, the long-tailed macaque, and an extinct elephant.[20] The reptiles were all species associated with water, crocodiles, gavials, turtles, and a monitor lizard. Storks, a goose, and a shelduck added to the reconstruction of an aquatic environment. Put together, *erectus* in Trinil lived near coastal rivers, lakes, swamp forests, lagoons, and marshes with minor marine influence. This diverse wetland system merged into grasslands on the drier ground. Trinil must have been an Eden for humans, perhaps more so than the drier environments of Africa where water would have been harder to find. Compared to their cousins in Africa, humans in Trinil were laughing.

The Trinil population seems to have been occupying habitats that had been used by the earliest humans on the island, who had entered 400 thousand years earlier during a period of lowered sea levels and found a low-relief lake margin landscape dominated by moist grasslands with open woodland in the driest areas.[21] This would not have been unfamiliar, a landscape dominated by water, trees, and open spaces. The climatic conditions, though, would have been different. In Africa, cooling and drying opened the landscape up to a greater extreme than in south-east Asia. Woodland became open savannah and grasslands became desert; water bodies were scattered. In the wetter world of south-east Asia, cooling and drying broke up the rainforest in favour of open woodland, savannahs, and grasslands but there was water everywhere. On top of all this the lowered sea levels, caused by global cooling and water being trapped as ice in polar regions, opened up land bridges between islands. While dry meant stress in Africa, in south-east Asia it meant paradise.

With subsequent warming and sea-level rise the humans became geographically isolated on these islands, which were sufficiently large to permit populations to persist in isolation for a long time. There were populations of the earliest humans (subspecies *erectus*) on Java as recently as 100 thousand years ago, and possibly even down to 50 thousand years ago.[22] With the resumption of rainfall the rainforest took over once more and the *erectus* population probably shrank in size and would have been confined to the coastal wetlands. But were they really trapped on these islands? If we follow the argument that I have used so far, it was always the marginal and stressed populations that were the innovators. Now, marginal and stress can mean different things to different populations depending on where they are, and when. For humans on south-east Asian islands warm and wet conditions were stressful because they reduced the area covered by open spaces. They lived on the margins of the great rainforest. So they would have been under pressure to find solutions for survival. Dispersal would have offered a way out on the mainland but here there was the open sea and no primate had ever crossed it.

Humans, it seems, did. The island of Flores, famous for the discovery in 2004 of the small hominid that became popularly known as the Hobbit, never had a land connection with either the mainland or any other island. So it was a real surprise when stone tools from Flores were dated to around 840 thousand years ago.[23] So how did humans get to Flores? We do not know. One answer would be that they had craft capable of travelling over water that far back but another would be that they drifted accidentally on logs and other floating debris between islands, particularly after storms. Some were able to establish a population by this means, just like many other animals, including monkeys, have done on countless occasions. What this mysterious finding tells us is that humans, who had lived along deltas and estuaries from around 1.8 million years ago onwards, were regular visitors to the coast;[24] so much so that they had been exposed to the possibility of, at the very least, occasional offshore drifting. Humans, from their very origins, were not only hominids of inland freshwater bodies but also of brackish and saltwater coastal ones. Our primate was even more improbable than we might have at first realized.

So it seems that humans in south-east Asia and in East Africa behaved in a similar way, living close to fresh water in areas with trees and open spaces, sometimes with rocky outcrops and caves. But they showed an ability to adapt, within these parameters, to differing circumstances afforded by accidents of geography. For example, they were much more coastal in south-east Asia than in Africa. For a long time we had no knowledge of *erectus* in the huge subcontinent that lay between south-east Asia and Africa but we now have new information that fills this gap in our knowledge. By at least 1.5 million years ago it seems that India was populated by people who were fully conversant with the Acheulian technology, including the manufacture of hand axes and cleavers.[25] In south-eastern India at least, a cluster of sites reveal the occupation of a coastal floodplain. Aquatic environments appear to have also been favourites among the first human populations in India.

From a similar time as the Indian sites, around 1.4 million years ago, humans were living along the shoreline of a wetland in today's Jordan Valley. The site of Ubeidiya was a mix of deep and shallow lakes, marshes, rivers, and wadis.[26] The mammal numbers were dominated by a hippopotamus but there were large numbers of deer, pigs, and a macaque which have been interpreted to represent woodland. My own experience of these species is that they are quite versatile and can also live in more open places, especially close to lakes and marshes. Even so there were also gazelles which indicate open patches of vegetation. There were large numbers of catfish as well as turtles and freshwater molluscs. Ubeidiya is another example of humans living in areas close to fresh water with trees and open spaces nearby at a very early stage. The environment they lived in, not surprisingly given the geographical proximity, resembled East Africa more than south-east Asia.

If we go west from Israel along the Mediterranean coast of North Africa we reach the sites of Ain Hanech and El-Kherba in north-east Algeria. These sites are dated to around 1.8 million years ago and represent an early presence of humans in North Africa.[27] At this early site humans seem to have been using only the early Oldowan technology. The site, really a complex of sites spread across an area of 1 square kilometre, has been described as an open and dry landscape which had a permanent body of water.[28] The sites produced numerous remains of hippopotamus, a species that would have needed year-round water to a depth of at least 1.5 metres, as well as crocodiles and aquatic turtles. The rest of the mammal fauna was dominated by horses, which indicated open, grassy environments where they could graze. The rich community of mammals included carnivores, elephants, antelopes, gazelles, giraffes, pigs, and rhinos. These animals are not forest dwellers, which is why an open landscape has been inferred. They are species, however, that are not restricted to open, treeless plains so it is very likely that wooded savannahs would also have been present. The North African sector of Middle Earth thus provided good conditions for humans who found the combination of habitat elements that seem to have defined the requirements of its lineage.

The earliest evidence from the northern shore of the Mediterranean, the northern fringes of Middle Earth, comes from the site of Sima de Elephante in Atapuerca, northern Spain,[29] and from sites around an ancient lake now disappeared, Lake Baza in south-east Spain, which have been dated to around 1.4–1.2 million years ago.[30] Whether humans got here 'the long way round', that is up through the Middle East and west across the Mediterranean coast of Europe, or instead managed to cross the Strait of Gibraltar from North Africa, is a matter of debate. In the absence of strong evidence for a sea crossing I prefer to stay with the long land route although we should keep in mind that humans (since I am regarding *erectus* as a subspecies of *Homo sapiens*) had somehow got to Flores before 840 thousand years ago. As always, we should keep our minds open to all possibilities. Whichever way they dispersed, they had reached this extreme corner of Europe by 1.4 million years ago.

As in Algeria, these early humans are associated with Oldowan technology only. The Acheulian portable kit did not arrive this early, or so we are told. These differences between sites and technologies have led to a variety of interpretations without satisfactory resolution. The answer may lie in the ecology. In Baza, as in Ain Hanech and south-east Asia, humans were living close to large and permanent wetlands so portable technology, needed when undertaking long journeys between water sources, may not have been important. Instead, they may have been happy with making more expeditious tools that served them well on the lake margins and which they did not have to carry over large distances. In tropical East Africa and India, seasonality in rainfall would have generated a greater patchwork of wetlands. In the permanent environments, the Oldowan would have sufficed but in the seasonal ones on the edge, the Acheulian came in handy. Viewed this way, it is easy to understand the co-occurrence of Oldowan and Acheulian in some places and the presence of one or the other in others. Both, according to this interpretation, were practised by humans depending on where they were and what they were doing.

This leaves one question unanswered, that of the late arrival of the Acheulian in Europe. The first recorded European Acheulian is from south-east Spain, not far from Lake Baza, from around 0.9 million years ago.[31] That is half-a-million years after the first humans arrived there. How can we explain the lag? I think that the conditions in western Europe, which have always been comparatively humid because of the proximity of the Atlantic Ocean, would have remained relatively wetter than in other parts of southern Middle Earth that had started to experience aridity. If the Acheulian was a response to dry conditions, providing portability when mobility was needed to find scattered water resources, then we would expect this technology to appear later in the more humid regions.

This is exactly what we find in Lake Baza. At its height it would have approached 60 kilometres in length and would have been another of the kind of wetland paradises in which we have already found the first humans. An impressive and diverse mammal fauna lived on its margins: rhinos, elephants, deer, bison, horses, sabre-toothed cats, large hyenas, and dogs. The lake itself held a population of hippos, and lots of frogs and toads have also been recovered, confirming the presence of fresh water. The list will be familiar to us by now, as I am sure it was for *Homo sapiens.* These humans probably got by without having to move across huge distances. Like Ain Hanech, this large lake would have held permanent water so the human strategy may have simply involved moving along the shoreline or into the lake shore itself at times when waters receded. That living conditions were favourable this far north is confirmed by a study of fossil species which are good climatic indicators: when humans were at Lake Baza mean annual temperatures were more than 4 °C higher than they are there today and mean annual rainfall was 400 millimetres over that of the present day.[32] The occupation of the edges of Middle Earth may have been discontinuous and dictated by climate, a topic that we will visit in Chapter 6.

Another site on the edge of Middle Earth that appears to have been occupied during mild conditions is Dmanisi in Georgia, which I referred to at the start of this chapter. The site was close to a river floodplain and had fauna which indicate open, savannah-like conditions with a range of species that included ostriches, deer, gazelles, elephants, rhinos, and horses along with jaguars, sabre-tooth cats, large hyenas, and bears. The stone tools were Oldowan.[33]

Further east and north-east, the earliest human presence in northern China dates to 1.66 million years ago,[34] and is therefore not much younger than the other sites that we have discussed so far. The oldest known site is Majuangou in the Nihewan Basin which, not unsurprisingly given what we have learnt so far about *erectus,* was a lake that alternated wetland and lake margin with a familiar fauna that included rhino, horse, elephant, deer, gazelle, a large hyena, and ostrich. The stone tools were of the Oldowan type. Even earlier is the human presence in the Yuanmou Basin in south-west China, around 1.7 million years ago.[35] These are the oldest humans on mainland East Asia and just younger than those on Java. Here they lived by a lake, made Oldowan tools, and lived alongside a fauna that was very similar to that in the Nihewan Basin and, indeed, many of the other sites that I have described.

When Dennell and Roebroeks[36] described their Savannahstan, they put grasslands as the common denominator that linked sites from the Atlantic coasts in the west all the way across to China in the east. This is understandable as we have seen how grazing animals were a feature of most early human sites. But the common denominator for me is fresh water. It is present in all sites and offered optimal conditions. The animals might have been different in Baza and Java but the water was the same. Wetlands, lakes, and river systems would have marked the dispersal pattern of humans as successive generations occupied similar places away from the centre of origin.

The proximity of first dates across Middle Earth is such that it makes it difficult for us to discern how and from where the early human populations spread but it is clear that it must have been a rapid dispersal. It seems that Europe—the Iberian Peninsula at least—and northern China were reached last but we have little to choose between East Africa, South Africa, North Africa, India, south China, or Java. My hunch is that it started somewhere in the core of Middle Earth. The edges of Middle Earth were occupied during conditions that were milder and wetter than today. Further south, seasonal rainfall and scattered water bodies were dealt with through increased mobility and the use of a portable stone-tool technology—the Acheulian. Exceptionally, in south-east Asia, rainforest expansion cornered humans in coastal wetlands and even led to inter-island dispersal. How south-east Asian sea crossings were achieved is a mystery which leaves the door ajar to other possible crossings of water bodies, like the Strait of Gibraltar. By 1 million years ago, humans had colonized all of Middle Earth. In Chapter 6 we will take a look at what happened next, when the climate took yet another downturn.

6

The Drying World of the Middle Pleistocene

800–400 THOUSAND YEARS AGO

The oscillating warm–cold cycles that characterized the northern hemisphere from ~800 thousand years ago are best known for the cold spikes that marked the start of the glaciations in northern Europe. These cycles were not just about temperature change. In the tropics the cycles are reflected instead by water, wet and dry periods being the dominant force of ecological change. The reality is that the climate oscillations of the Middle Pleistocene (the geological period ~780–125 thousand years ago) generated severe phases of aridity across the world, not just the tropics. These cycles introduced desertification across large areas of Middle Earth[1] and they were responsible for the geographical expansion and contraction of ranges, and the extinction of many species. The important consequence in the context of our story is that aridity generated ecological barriers between different regions of Middle Earth. The result was isolation of different regional human populations.

A major step that marked a significant climate change in East Africa around 1 million years ago took the continent another step towards drying up. This was not a one-way ticket towards aridity but took, instead, the form of violent swings in climate which heightened the extremes.[2] The droughts became more severe and alternated with wetter moments. African human populations in this period would have felt that their world was in turmoil. Aridification must have affected adjacent regions too, especially North and South Africa, the Middle East, Arabia, and the Indian subcontinent. The paradox is that between 1 million and 400 thousand years ago the African human populations seem to spread out and into drier and cooler environments.[3] In South Africa, for example, humans with Acheulian technology had survived only in the wetter north and east prior to 1 million years ago but after this point spread into the drier interior and the western margins. How can we explain the apparent paradox?

We can do so very easily in my opinion. In Chapter 5 I suggested that the makers of the Oldowan tools had tended to occupy areas next to permanent water bodies. In other words, they lived in fairly stable environments compared to more unstable places where they would have had to move around over large distances in search of water. The Acheulian gave them portability in such circumstances. As the world dried further, those populations able to make Acheulian as well as Oldowan tools fared better than those who could only make Oldowan. Acheulian people were the ones that could deal with variable and arid conditions. They had got used to these conditions so that, when the climate became drier and more variable after 1 million

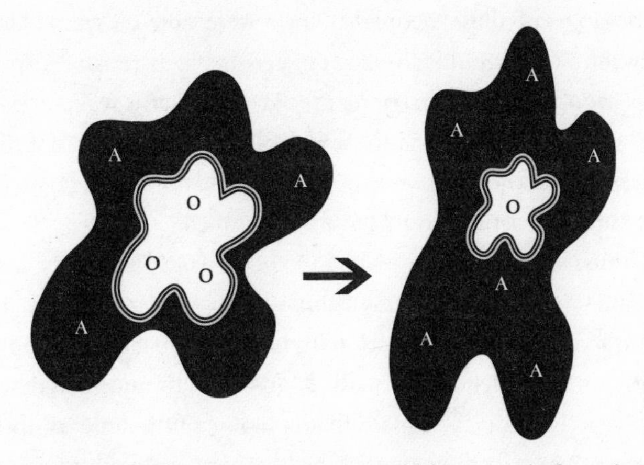

FIGURE 4. How technologies can coexist, survive, and disappear. The figure shows a first scenario in which Oldowan (O) and Acheulian (A) coexist in a humid environment. The Acheulian exists only on the drier, peripheral environments where its portability allows its makers mobility in search of scattered water resources. In the second scenario, climate has become more arid and the Acheulian dominates large areas, the Oldowan becoming severely restricted.

years ago, they were able to cope and even expand into arid environments. We would then expect that, as the world became more arid the Oldowan would disappear altogether but the Acheulian would remain and that is exactly what happened: after more than 1 million years of the two technologies, the Oldowan fizzled out during the increasingly arid conditions of the Middle Pleistocene[4] (Fig. 4).

In South Africa, the Middle Pleistocene human populations with Acheulian moved across the landscape with reference to available surface water (springs, pans, rivers) and stone outcrops where tools could be made.[5] Acheulian tools were made precisely to be carried, giving some degree of independence from stone quarries, but we have no knowledge of the water-carrying technology of these people. If they were able to carry water between sites it would have been in perishable materials—ostrich eggs, wooden containers, or plants that could act as receptacles—that would be invisible in the archaeological record. Even if they did, return to known sources of water would have been regular as any supplies carried would have been limited.

Humans lived at Olorgesailie in the Kenyan Rift Valley from around 990 to around 625 thousand years ago, roughly 350 thousand years of occupation that was probably interrupted many times by depopulation. It is important to understand how coarse archaeological signals, that could be of the order of thousands of years, cannot be matched to human generation times which are on a completely different scale of time. At Olorgesailie archaeological horizons are thought to represent brief periods of around 1 thousand years[6] and conclusions have been drawn about human behaviour on the basis of comparisons of different strata. There is always a risk in assuming that such long periods were homogeneous, which is extremely unlikely.

Nevertheless, Olorgesailie does give us very useful insights. The useful information is that the combination of water, trees, and open spaces is applicable throughout the long sequence. Rocky places were also a regular feature, at least as sources of raw materials with which to make tools. Whenever people were at Olorgesailie, it was in that context. It seems that sometimes conditions were drier and other times wetter—a lake expanded and shrank in size, streams sometimes flowed freely with ample water and other times were reduced to trickles—and the distribution of stone tools on the landscape changed. So humans may have been responding to changes in the proportion and distribution of trees, open spaces, and water. One suggestion has been that these people lived in the hills, where the raw materials for making their tools were available, and they descended to the lake with these tools in search of food and water. This would imply that humans fed directly where they caught or scavenged large mammals or, instead, took chunks of meat and large bones filled with marrow back home. That seems fine, but what about water? This interpretation gives us two options: they only drank when they were by the source of water and somehow managed to spend the rest of the time without drinking or they had containers for carrying water. Of course, we can interpret the observations in other ways. One would be that *erectus* went to the higher ground in search of stone and had its base either close to water or, if that was too dangerous because of the risk of predators close by, somewhere in between and close to both rocks and water. What Olorgesailie does also confirm very clearly is the mobility of people who, by this time, were introducing some stone artefacts from sources that were a minimum of 26–45 kilometres away.

Around 780 thousand years ago, a community of humans was living by a lake that was not far from Ubeidiya (Chapter 5). The site of Gesher Benot Ya'aqov, usually referred to as GBY, is in the northern part of the Dead Sea Rift, some 12 kilometres north of the Sea of Galilee. It has become emblematic for a variety of reasons, most importantly because it has the earliest evidence for the processing of plant foods and the working of wood.[7] The people who lived at GBY were always close to water, sometimes a river and at other times a lake, with a landscape composed of trees and open spaces judging from the plants recovered and also the presence of the straight-tusked elephant which was apparently hunted and butchered on site.[8] The archaeological site is waterlogged and this accident of Nature has preserved a large range of plant material that would otherwise have deteriorated. The range of plant materials included wood, bark, fruit, and seeds. Many were edible and a large number were plants that live immersed in or float on fresh water.[9] Fifty-four stones were recovered from the site. These were pitted in ways that indicate that they had been used as hammers and anvils to crack nuts of seven different species, including two aquatic ones.[10] Burnt seeds, wood, and flint found at GBY provide the evidence indicating that humans at this site had control over fire.[11] The uneven pattern of distribution of burnt fragments of flint suggests that burning was localized in particular areas that could be interpreted to have been hearths. The wood of several plants—including the edible olive, wild barley, and grape—was found burnt, providing a possible link between fires and cooking.

GBY represents the earliest known control and use of fire but Richard Wrangham at Harvard University argues that it is much older and was a defining feature of the first humans.[12] The evidence, Wrangham argues, comes from our anatomy. With humans of 1.8 million years ago we see a reduction in jaw size, brain enlargement, and a shortened gut. These anatomical changes were possible because humans were cooking their food, making it easier to chew and digest. Energy was extracted more efficiently from cooked food which helped to fuel a large brain. The idea is appealing, and may well be close to reality, but it will always suffer from the retort challenging the lack of direct evidence of the hearths that humans made nearly 2 million years ago. Apart from its use in cooking, fire would have served additional functions, such as keeping predators at bay at night and providing warmth, something that would have been particularly important in the cooling environment of the temperate parts of the world during the Middle Pleistocene. We should remain open to the possibility that humans controlled fire from an early stage. If they had it at GBY at 780 thousand years ago, the chances are that its use started some time before. Just how much earlier is hard to pinpoint.

Middle Pleistocene sites with the quality of preservation and depth of occupation of GBY and Olorgesailie are rare. Robin Dennell has reviewed the Asian sites[13] and there is very little that can compare. A common denominator of sites across Arabia, India, China, and south-east Asia is water as most sites are close to rivers, lakes, or other wetlands. Evidence of living on the coast is harder to find but this must reflect a problem related to loss of coastal archaeological sites because of rising sea levels. Middle Pleistocene populations would have lived on the coast and exploited its rich resources, judging from the evidence from south-east Asia (Chapter 5). Europe, Asia's westernmost peninsula, has been intensely studied since the 19th century and has a few sites that offer comparable levels of detail to GBY and Olorgesailie. Clive Gamble's reviews are still useful summaries of the European Middle Pleistocene picture.[14] The majority of European Middle Pleistocene sites are riverside and lakeside gatherings of humans where they seem to be butchering large mammals. These mammals are typical species of European open woodland, of places that combined trees and open spaces.[15] The picture across Middle Pleistocene Eurasia is one in which humans lived in places close to fresh water, with trees and open spaces, and sometimes in rocky areas where they seem to have started the habit of entering and living in caves, even if only sporadically.

The pattern of human occupation of Eurasia was heavily dependent on rainfall, temperature, and—in northern latitudes—also day length. In Chapter 5 we established that the first confirmed human presence in western Europe was in the Iberian Peninsula around 1.4 million years ago. We do not know if that date marked the establishment of a permanent human presence in Europe or if, instead, European populations repeatedly went extinct locally and were replenished by later immigrants. It would take them another half-a-million years to reach higher latitudes in Europe. The earliest recorded presence in north-western Europe is around 700 thousand years ago, at Pakefield in Suffolk which is at 52°N.[16] Curiously, this date is after the first known use of controlled fire by humans which poses the intriguing question whether fire was the key to the expansion to higher latitudes at a time when the world was getting cooler and in areas where long winter nights would have curtailed activities normally carried out in daylight. Having said this, the presence of people at 52°N coincided with an interglacial, a warm period sandwiched between colder episodes. This trend, of a presence in the north only during warm periods, seems to be prevalent across Middle Pleistocene central and northern Europe, borne out by the long periods when Britain was deserted.[17]

The human presence along the northern fringes of Middle Earth, not just on the European peninsula, during the Middle Pleistocene was limited to warm periods between glacials.[18] Central Asia was abandoned during cold and dry periods and the growing central Asian deserts became a significant barrier. China, especially the north, seems to have been occupied only during interglacials and interstadials. A similar trend is observed along the higher ground, in the mountains of Middle Earth that replicated the effects of latitude across short vertical distances. Caves were occupied sporadically in the Caucasus and the central Asian mountains. One of the better known sites is the huge cave of Sel'ungur in Kyrgyzstan which is situated at 2,000 metres above sea level and was occupied by people who left the cave during cold periods. When living in this cave they had a panoramic vista outside that included mountain steppe, desert, areas with trees and shrubs by ephemeral streams within wet depressions.[19] It seems that the combination trees/open-spaces/water, here combined with rocky habitats, also worked for humans high up in the mountains of central Asia.

In the southern part of Middle Earth, in North Africa, Arabia, India, and south-east Asia, day length and temperature would not have had the impact that they had in the higher latitudes. Here it was shortages of rainfall that caused havoc. At Hunsgi-Baichbal in southern India,[20] humans lived close to rivers and streams during the wet monsoon but aggregated around shrinking water sources in the dry season. Then they seem to have hunted medium to large game that was attracted to the few remaining water sources. In the wet season groups were more widely dispersed across the landscape, when many smaller streams carried water. Humans then had a wider range of resources available that included many edible plants and small game. In the Arabian Peninsula, today engulfed by the desert, Acheulian sites were associated with a range of environments, including coastal zones, high ground, and interior plains, occurring along river terraces and near lake shores.[21] This occupation of Arabia would have happened during wet cycles when desert gave way to steppe, grassland, and savannah.

An intriguing possibility, and a complete contrast to India or Arabia, has been offered as a response of people to drying conditions in the humid tropical contexts of south-east Asia. Panxian

Dadong in south-east China is situated at 1,630 metres above sea level. Here humans lived among pandas, on a valley floor at the confluence of three rivers in a landscape of mixed woodland, bamboo forest, and open rocky areas. It seems that humans colonized these uplands during cold and dry periods when water shortages forced populations to move to higher ground, which was wetter.[22] Cooler conditions would not have had the impact in the tropics that they had in higher latitudes so moving up mountains would not have been restricted by temperature.

When I was starting to write about human evolution, I wrote a paper with my wife Geraldine and my friend and colleague at the Gibraltar Museum, Darren Fa. The paper was published in 2000 and it summarized our views on human geography at the time.[23] The three of us were ecologists and biogeographers and we were unsatisfied with the existing models of human dispersal. In that paper, and again later in my book *Neanderthals and Modern Humans,*[24] we put forward a model of sources and sinks in human evolution. In it we proposed that the persistence of populations of *Homo* (including *erectus*) was expected to be highest in tropical areas, which we defined as sources, and lowest in higher latitudes, which we defined as sinks.[25] Tropical Africa was a major source area and we argued that south-east Asia became a source area once colonized. At the time we were working with the established paradigm that *Homo erectus* had colonized Asia from Africa. We argued that mid-latitude areas, such as the Mediterranean, were secondary sources where populations could sometimes hang on during bad times, which fed populations at higher latitudes. This model fits with the idea that populations on the northern edge of Middle Earth kept going locally extinct and subsequently restocked from the south.

There is another, very important, factor that we need to add to this north–south interpretation of human survival based on temperature and day length. That factor is water, and its impact was huge at a time when the world was becoming much drier. In particular, the expansion of the central Asian deserts and the Sahara introduced major barriers to the movement of the water-dependent humans. Like Arabia, the Sahara alternated from barren desert to a land of lakes[26] when humans would have thrived along lakesides and riversides just as *Sahelanthropus* and the australopithecines had done millions of years earlier. The Sahara was part-time paradise and part-time hell. To understand the implications of these gigantic desert barriers I will draw from the world of birds. If we compare the diversity of birds in the west and the east of Eurasia, Europe, and China respectively, we find that China wins easily.[27] Species of tropical families are also better represented in China than in Europe and we can find similar patterns among other groups. Part of the reason for the difference, at least, lies in the climatic continuity between China and tropical regions to the south while Europe is severed from tropical regions to the south by the Sahara Desert. This means that extinction would have been higher in Europe than in China during glacial periods. In Europe populations became trapped in the Mediterranean cul-de-sac but in China they could shift their ranges further south. Replenishment from the south during warmer periods was also easier in China than in Europe. One consequence, in birds at least, is that the European fauna has stronger affinities with the east than with the immediate south.

In the Middle Pleistocene, the Saharan, Arabian, and central Asian deserts had the combined impact of disconnecting the human populations of tropical Africa from Europe and both populations from those of the Far East. The absence of water was contributing to a biogeographical pattern of isolation. If to this we add the effect of the glaciations in the northern hemisphere we have to reach the conclusion that the populations that fared worst must have been those from Europe and adjacent regions because they were repeatedly pegged back by glacial conditions and were kept isolated from populations to the south and east by the great deserts. The populations that fared best would have probably been those of eastern Asia. Cold and arid conditions in the north made it unliveable, as we have seen, but they were countered by the opening up of rainforest further south and this generated favourable conditions for humans. The cold conditions also lowered sea levels and exposed additional land that could be colonized or that could act as bridges to other land masses. Tropical Africa had the benefit of latitude which meant that the full impact of the cold did not compare with the higher latitudes but it was also less wet than south-east Asia so dry periods intensified desertification and reduced the surface area of habitable land. India would have had connection with south-east Asia during dry periods, when the rainforests opened up, and with Arabia and the Middle East during wet periods, but the Himalayan wall to the north always prevented north–south movement.

From 600 thousand years ago onwards, many authors refer to the living *Homo* populations as belonging to a new species, *Homo heidelbergensis.*[28] The main reason is a perceived step in brain size at this time which has been considered to represent a speciation event, the formation of a new species of hominid. According to this view, *Homo erectus* continued in east and south-east Asia and *heidelbergensis* occupied western and central Eurasia and Africa. Others prefer to separate the Eurasian and African populations, *heidelbergensis* in Eurasia and *rhodesiensis* in Africa. On the one hand it seems clear that connections between intercontinental populations of humans became severed during the Middle Pleistocene. The fragmentation of intracontinental populations also seems to have been important, especially in western and central Eurasia. This repeated segregation would have promoted differences between populations through genetic isolation so it should not surprise us that some researchers have even tried to turn Eurasian *heidelbergensis* into several species. I am taking a step in the opposite direction by considering all populations after 1.8 million years ago, including the Middle Pleistocene ones, as belonging to *Homo sapiens.*

The reasons are very clear to me. On the one hand, I would argue that the supposed sharp increase in brain size at 600 thousand years ago is not real but an artefact of the small samples available to us. We can explain the increase in brain volume, instead, as one of exponential growth, which would appear as a practically imperceptible change in size for a long time followed by a sharp increase. So there would then be no sharp break that would allow us to separate *heidelbergensis* from *erectus.* What about intercontinental isolation? The timing of the split between modern human and Neanderthal lineages coincides with the Middle Pleistocene separation of western Eurasia from Africa and the two lineages remained separate for around half-a-million years. Yet, we now know that when they met they exchanged genes and therefore behaved as members of one species. If half-a-million years of isolation were insufficient to turn modern humans and Neanderthals into separate species, then it is highly improbable that a shorter time of isolation between populations of *Homo sapiens* (or *erectus* or *heidelbergensis* as others would prefer) would have achieved that result.

After a million years of occupation of Middle Earth under comparatively mild conditions, the drying world of the Middle Pleistocene, particularly after 800 thousand years ago, was responsible for the breakup of populations between continental land masses and within continents too. For a species that had become intimately tied to water, aridity spelt disaster. We should not forget that the Middle Pleistocene was a world of climatic swings. These breakups were mitigated, in some places at least, by reinvasion of lost territory when conditions improved. Reinvasion would have followed the linear watery world of rivers, punctuated by the bonanzas provided by lakes, swamps, and marshes. Such reinvasions, many of which we may not have the resolution to pick up with the archaeology, would have remixed genes. The result was that the unity of *Homo sapiens* as a polytpic species[29] was maintained throughout.

7

The Rain Chasers— Solutions in a Drying World

400 TO 70 THOUSAND YEARS AGO

Homo erectus, or as interpreted here, *Homo sapiens erectus*, had the beginnings of a body built for endurance running and walking, a body that would become increasingly refined in some populations as time went by.[1] Some human populations sacrificed muscular power for lightness that enabled an efficient way of getting across large tracts of open landscapes. Other than the oft-quoted advantages in the pursuit of prey or in the early arrival at carcasses, lightness of build would have provided a key advantage in a water-limited world. This advantage facilitated the seeking-out of ephemeral and widely scattered sources of water, the tracking of seasonal flushes of grass, and the gatherings of herbivorous mammals that followed. This is the basis of the rain chasers, the human populations living in southern Middle Earth that increasingly adapted to cope with the opening-up of the landscape as water became an increasingly limiting and limited resource.

In Chapter 6 I argued against a speciation event around 600 thousand years ago based on a rapid increase in brain size, proposing instead a steady and gradual increase in brain size in *Homo sapiens* populations. As this evolution was gradual I could see no case in favour of renaming humans after 600 thousand years ago. Now let us move forward to seek out human populations in southern Middle Earth, after 400 thousand years ago, when the world's climate had deteriorated even further than before ~450 thousand years ago. The 100-thousand-year cycles of warm and cold, that started around 800 thousand years ago, continued but with important differences. The extremes became greater than before and the length of the warmest intervals shortened. The changes in climate also became more abrupt, with transitions between cold and warm being very rapid. In Africa, the cold–warm swings were translated into wet–dry oscillations. Dry periods became increasingly prevalent. Africa was drying up even more quickly than before.[2]

These climate changes not only isolated humans in different continents, as we saw in Chapter 6, but they also cut off populations within continents. In southern Middle Earth the fragmentation was driven by water shortage and these isolated populations experienced severe genetic bottlenecks[3] which in the worst moments reduced numbers to a few thousand individuals. My argument here is that it was the shortage and scattered nature of water resources that drove natural selection to favour individuals that could best cope in this kind of water-limited world. The period after 450 thousand years ago was no different from the preceding 1.5 million years in that respect, except that the severity of the climatic conditions was far worse. The human populations simply had to do what they had been doing for a long time but each time they had to get better at doing so as the climatic signals got stronger.

What did these *sapiens* living in southern Middle Earth after 450 thousand years ago look like? We have a reasonably good image of their appearance, which contrasts with that of the Neanderthals who were their contemporaries in Eurasia (Chapter 8). The *sapiens* had long legs, a linear physique, narrow pelvis, and low body mass for their stature.[4] In other words they were slender, streamlined, and lightweight. Their anatomy, including the well-developed Achilles tendon, indicates that they were endurance runners and long-distance walkers.[5] The humans of after 450 thousand years ago were fine-tuned versions of the earlier models, all of whom from their origins 1.8 million years ago had been selected for the ability to cover huge areas of ground quickly on two legs. We have also seen how large brains would have been useful hard disks that stored masses of data accumulated during the life of each individual person. These hard disks also had a good RAM, which allowed easy retrieval of the stored information. This meant that sources of water could be picked up in a dry landscape from memory, just as the earliest ancestors had retained information about where and when fruiting trees were in season. The same ability was working in favour of humans in southern Middle Earth. Not only was the post-cranial anatomy constantly being tweaked for improved performance but the cranium and its soft contents were also the subject of focused attention by natural selection. The fashion was towards an increasingly slender and large-brained human and this trend was continuous through time. It did not, in my view, involve any sudden step, implying the emergence of a new species, as has been repeatedly suggested.

In the same way that the arrival of the new *Homo heidelbergensis* has been claimed for around 600 thousand years ago, so the entrance of *Homo sapiens* is marked at around 200 thousand years ago by the oldest known fossil with characteristics of our species.[6] Yet, when we look at *Homo sapiens sapiens,* especially at these initial stages of its career, we tend to find a hominid that resembles us but that also retains many features from its ancestors. This has led palaeoanthropologists to introduce such paradoxical terms as 'archaic modern humans' to define these populations. They exhibit a mosaic of features that illustrates that we are observing a gradually evolving lineage. In the drying world of Africa and adjacent regions of southern Middle Earth after 450 thousand years ago, populations were becoming isolated and were then reuniting as climatic conditions changed. In this situation of instability, isolated populations may have developed distinctive anatomical features which may have subsequently become intermixed with those of other populations when they reunited, creating the impression of the evolution of a medley of characters.

It is easy to fall into the trap of looking at ourselves today and the first populations (*H. s. erectus*) and seeing the differences that convince us that we are looking at two different species. It is quite possible that they are linked by intermediate forms which unite the lineage, just as intermediates unite geographically disparate forms in a polytypic species (see Chapter 6 note 29). This interpretation has been championed by Milford Wolpoff at the University of Michigan who has argued in favour of 'a geographically dispersed polytypic species, *Homo erectus,* evolving into a geographically dispersed polytypic species, *Homo sapiens*' and that 'Taxonomically, this means that there is but one species of *Homo*, *Homo sapiens*'.[7] This does not mean that there was no branching of separate lineages during the long course of our evolution. A case in point is the divergence of lineages that led to the Neanderthals and an African family, which is the one we are discussing in this chapter. These lineages probably spawned many other lines that had their own genetic identity, without necessarily achieving the distinctness that would warrant taxonomic separation as species.[8]

The southern Middle Earth—northern Africa, Arabia, and India—human populations reveal a great capacity for technological innovation, which is hardly surprising given their large brains and the stressful conditions in which they were evolving. It was in these marginal populations that the pressure to invent new ways of dealing with their environment was greatest. Technological innovation would have started with the first stone tools and was later refined with the invention of the Acheulian technology. The new pressures after 450 thousand years ago stimulated further novelties.

Technological fashion appears to match the anatomical trend, which is towards the invention of lightweight and portable tools. This makes sense if the pressures on people in the drying world of southern Middle Earth were towards larger and larger home ranges, rapid location of highly dispersed and ephemeral sources of water, and the ability to track rain fronts. The last would be observable at long range on the African plains and humans would have been keen to reach these quickly, not just for the water but also because the flushes of grass that would result from the downpours would attract the herds of migrating herbivores. Rain meant food as well as water. In these situations, the rain might not fall where there were sources of stone to make tools from so it was best to carry the stone to the water. Acheulian tools were portable and people continued to use this technology for a long time but, as new techniques were invented, they were probably used alongside the Acheulian or even in preference to the tried-and-tested ancient technology. There is an obvious parallel with the earlier discussed use of Oldowan and Acheulian tools.

The most noteworthy novelty that post-dated the Acheulian was the invention of hafted tools and the production of small flakes from previously prepared stone cores.[9] The earliest evidence of this new technology dates to around 300 thousand years ago but there are glimpses, requiring further study, which might take the start back further, possibly beyond 340 thousand years ago. This has been interpreted to mean that this new technology pre-dated *sapiens* by at least 100 thousand years and would have been invented by *erectus* or *heidelbergensis*, considered as separate species. The lack of correlation between the different perceived species of *Homo* and technological novelty has been a source of concern for archaeologists who, in order to explain the mess, have even suggested that different species of *Homo* might have coexisted.

If we adopt the view that there was only a single species—*Homo sapiens*—then the problem is instantly resolved. What we have are groups of large-brained, intelligent *Homo sapiens* capable of improvising and inventing new solutions for dealing with their environment and transmitting this information to other groups. Invention has always been a cocktail of intelligence, improvisation, need, and chance, so, for a large-brained hominid, new ways of dealing with the world could have come up anywhere and at any point, which is what we find in the archaeological record.[10] Technology in humans owes its existence to the pressures of finding dispersed sources of fruit in the forest, later water in a drying world, coupled with the need to maintain complex social relationships within and between groups.[11]

Hafted technology conferred huge advantages to its nomadic makers. It was economical on stone as more flakes could be extracted from a core than with the Acheulian tools; the haft could be reused when the stone point was spent, by simply replacing it; energy was saved as less force was required by the leverage provided by the extended arm of the haft; and it reduced the risk of injury to the bearer by putting distance between hunter and hunted. Hafted tools could also be carried more easily because they were lightweight in comparison to the Acheulian hand axes, so they made the life of the plains-wanderer easier all round.

Composite technology, as in hafting, in which the tool or weapon was composed from distinct items, meant that humans could respond quickly to rapidly shifting resources such as water and migrating herds of herbivores. The improvements to mobility that were triggered by having to move between water sources were now coming to the fore in the context of chasing highly mobile animals; this ability to hunt mobile prey was a by-product, I would argue, of adaptations resulting from the need to search for highly dispersed water. The new technology was especially useful in highly seasonal environments where food was constantly on the move; with it humans could make specialized tools that allowed them to process abundant and short-lived resources such as fish, and also to target large, wide-ranging, and potentially dangerous prey.

The composite tools, including hafted ones, were probably multi-purpose, like Swiss Army knives. A tool could have served a range of jobs. Such multiplicity of function would have served to minimize risk when living in a rapidly changing environment or when entering new and unfamiliar territory by reducing the possibility of being ill-equipped. So, overall, what technology is showing us is a parallel development of ways of dealing with a drying world in which resources were increasingly dispersed and hard to find and where travelling light was essential in order to cover large distances. This was all part of a gradual process that started with the first humans 1.8 million years ago. What technology did was to add a fits-and-starts element to the gradual process. Inventions happened when they happened, spontaneously and not gradually. Fashions spread. If new tools and weapons were useful and there was connectivity between human populations the novelties would spread quickly. Importantly, if there was connectivity with other lineages, for example the Neanderthals to the north, then the novelty could work its way geographically into these populations too (see Chapter 8). In turn, technology would have influenced the behaviour of humans by allowing them to put themselves in situations that they could not have been in previously. The resultant lifestyles in turn impacted on the evolution of human brains and bodies.

If we take a look at what was happening across different parts of southern Middle Earth during this drying after 450 thousand years ago, we find evidence of people living close to water and of movement across large distances. We also observe behaviours that have been claimed to be novel—such as the use of the coast and its resources—but which I will question in the brief review that follows. Some regions became more arid than others. Among the worst affected was North Africa which suffered from the expansion of the Sahara Desert.[12] The Nubians[13] lived along the banks of the Nile and the lakesides of the eastern Sahara where they butchered a range of herbivores including gazelles, rhinos, giraffes, and water buffalo. They would have been exploiting habitats close to water that included grassy, open spaces, and some trees, judging from the animals exploited. They also ventured into the mountains of the Red Sea coast where they hunted buffalo as well as elephant and kudu,[14] seemingly tracking similar environments onto the higher ground where there was likely to be greater rainfall, in a similar fashion to earlier peoples in south-east Asia (Chapter 6). It seems that as conditions got drier human occupation became sporadic except in the eastern Sahara where they remained close to lakes until the onset of extreme aridity after 77 thousand years ago.

Where geography brought together very different worlds, humans made the most of the diversity. At Haua Fteah in present-day Libya, they went inland in search of gazelles and wild cattle in open areas with trees but they also entered drier, rockier areas where they hunted Barbary sheep.[15] The exploitation of gazelles and Barbary sheep seems to have alternated with availability, the former in wetter conditions and the latter during drier moments, showing the adaptability of humans to aridity in these extreme situations close to the Sahara. This adaptability is also reflected in the exploitation of coastal shellfish, taking advantage of the site's location.[16] At Taforalt in Morocco, humans collected shells from the coast which was 40 kilometres away and at Oued Djebanna in Algeria a shell must have travelled a distance of 200 kilometres.[17] These observations are important, not because they might represent the sudden arrival of modern human behaviour,[18] but because they hint at the degree to which humans had been able to extend their mobility by 82 thousand years ago. These North Africans may have had large annual home ranges that included coastal and inland areas or they may have established trading networks with neighbouring groups, or both.

The perception that humans suddenly discovered the coast around 170 to 160 thousand years ago[19] stems from the idea that the exploitation of coastal sites in South Africa at this time was a major event that marked humanity's arrival on the scene and its subsequent global expansion along coastal routes. The coast was probably exploited from very early times by people and certainly by 800 thousand years ago, judging from the south-east Asian evidence which I discussed in Chapter 5. In Chapter 8 we will find the Neanderthals on the coast too. So there is no big deal about finding humans living on the coast. We simply find more evidence of it after 125 thousand years ago because the sea-level rise during the global warming event around that time washed away much of the earlier evidence of coastal habitation. Pinnacle Point in South Africa, the site with the earliest claimed date, was exceptional because the lie of the land (steep cliffs) protected it from that particular sea-level rise. The evidence of earlier coastal exploitation, if it exists, must be sought in the seabed.

What the South African sites do show is that humans regularly lived in caves, probably using these for shelter as the climate became cooler. We have seen how the use of rocky habitats was a feature of some human populations from the outset and it is possible that such places were visited more frequently as the climate became colder and more arid. Rocky places would have been substitutes for trees where these were disappearing because of drought. Here people could have hidden to ambush Barbary sheep and ibexes, just as they ambushed gazelles in open woodland. By the sea they would have ambushed breeding seals and seabirds and collected shellfish typical of rocky coasts.[20] In the absence of trees, rocky areas would have provided places in which to sleep at night safe from predators, especially with entrances protected by fire. Away from the tropics, such as in North and South Africa where the effects of the cold would have been most felt, caves would have provided additional incentives.

The South African sites also show a connection with fresh water, being located on or close to coastal estuaries which would have had the added benefit of a rich resource base. Coastal sites, away from estuaries and deltas, were probably also sought-after as places where fresh water could be found. Many areas of the continental shelf, now submerged, would have offered coastal oases[21] to humans. These would have been places where rain-water percolated from higher ground and surfaced. During cold periods, when sea level dropped, the removal of the pressure of the sea above these springs permitted fresh water to surface with greater force and in greater quantities, creating chains of coastal oases along the emerged coastal shelf.

In Aduma in Ethiopia, humans were producing even smaller tools than before, presumably in response to further aridity, and there is evidence of long-distance transport.[22] The situation resembles that in North Africa and humans here used a variety of habitats, some of which were close to, and others away from, a river. This would seem to suggest that humans were each time venturing over greater distances across arid landscapes. Food, like water, was obtained close to the river, judging from the presence of hippos, crocodiles, and large catfish at this site. Water remained a magnet which tells us that, like in the eastern Saharan lakes, human populations were tied to surviving water bodies in times of extreme drought and that mobility had helped but had not granted them independence. The Mumbwa Caves in central Africa were abandoned when dry conditions set in and the area surrounding them was less attractive to animals which relied on seasonal standing water and which were hunted by people.[23] This example shows that, when conditions were really bad, people faced death or rapid emigration.

The period between 450 and 70 thousand years ago—the latter marking the onset of the latest and most intense glaciation—was one in which continental Africa experienced severe climatic fluctuations and, importantly, drought. The changes were expressed not just in the distribution and abundance of supplies of water, but also in the further opening-up of the vegetation and the expansion of treeless environments: grasslands, steppe, and desert. The human populations of southern Middle Earth suffered heavy losses. We see that at the local scale in the abandonment of sites and concentration of populations close to the remaining water bodies; we see it at continental scale through genetic bottlenecks that brought numbers down to a few thousand.

These huge pressures fine-tuned the survivors in the direction of large brains and lightweight bodies capable of covering lots of ground quickly in search of water. These human populations became the ultimate rain chasers. The trend towards economical movement and transport went beyond the purely biological and into the realm of technology and culture. People made increasingly light, economical, and recyclable weapons and tools. The solution that just, and only just, allowed humans to survive in southern Middle Earth, in an uncertain world with highly dispersed critical resources, was mobility and risk reduction. Let us now turn to Eurasia and see the response there of *Homo sapiens* to a different set of conditions.

8

The Exceptional World of the Neanderthal

400–30 THOUSAND YEARS AGO

As *Homo sapiens sapiens*—the lineage of southern Middle Earth—was starting to feel the pinch after 450 thousand years ago, a separate lineage had embarked on its own particular adventure in the non-tropical regions of Eurasia. That lineage had a common ancestor with *sapiens*, which genetic work has put at around 400 thousand years ago.[1] This Eurasian lineage is often referred to as *Homo heidelbergensis*, the same name that is applied to the African lineage of *Homo sapiens* from the same period and which was the subject of Chapter 7. Others, as mentioned earlier, distinguish the African and Eurasian lineages, giving the name *rhodesiensis* to the Africans and retaining *heidelbergensis* for the Eurasian contingent. The nomenclature has clouded the continuous evolutionary process. At one end of the scale some authors have tried to make each regional version of *heidelbergensis* into a species in its own right and at the other end they have merged them into one. When considering fossils that show some characteristics typical of the Neanderthals, some authors refer to these as *heidelbergensis* while others talk of pre-Neanderthals, a term that reminds me of 'archaic modern human' and reveals the inadequacy of trying to put a continuous evolutionary process in a straitjacket. Since this lineage led to the Neanderthals but was derived from the common *sapiens* ancestor, I shall refer to it as *Homo sapiens neanderthalensis.* For our purposes they are the Neanderthals, even if the earliest ones do not have all the features that have been defined as characteristic of them.

The Neanderthals lived across Eurasia but we cannot determine where they originated with exactitude, although the common but unsupported view is that it was in western Europe.[2] At the height of their success, their range extended from the Iberian Peninsula in the west right across to central Siberia, and probably beyond, in the east.[3] The achievement of the Neanderthals should not be underestimated and it matches that of humans in the drying world of southern Middle Earth. In Chapter 6 we saw how humans had struggled to survive on the northern edge of their range, in Europe, central Asia, and China. Occupation of northern territories was not permanent and populations were pegged back during cold periods. If that was bad, imagine what followed after 450 thousand years ago. The conditions that brought severe drought to Africa also brought extreme cold to Eurasia.

The Neanderthals, like their predecessors, could only survive in warm and humid refuges each time that the ice set off on a southerly excursion. The coastal south-westernmost tip of the Iberian Peninsula between Gibraltar and Lisbon, the mildest and least arid part of Europe, was their main stronghold. The limit of their geographical range was marked in the south by the Mediterranean Sea and its high mountains. High mountains ran from there across Middle Earth, all the way to China. They were formidable barriers which were impenetrable at the best of times, let alone during a glaciation when ice descended from the lofty heights. Each time this happened the Neanderthals were trapped between frozen northern latitudes and icy southern mountains but they picked themselves up during warm interglacials and spread northwards and upwards once more, only to be pushed back once again with the next climatic downturn.

A little while back I decided to take a look at the geographical distribution of the Neanderthals during the last time that their range contracted.[4] It was significant because the Neanderthals would never recover from it. I looked at the period between 50 and 30 thousand years ago, the last known Neanderthals having survived to around 31.5 thousand years ago at Gorham's Cave in Gibraltar.[5] Using a range of criteria of suitability of different regions to the Neanderthals I predicted the extinction date for regions of Europe and the Middle East and found that the predicted last survival dates matched closely actual published dates. The results showed that the process of population extinction had been long drawn out rather than a rapid event. This contrasted with the view that around 40 thousand years ago the rapid arrival of modern humans into Europe had generated a kind of blitzkrieg on the Neanderthals with a rapid extinction across Europe.[6]

I found that the pattern of extinction across Europe and the Middle East had not been uniform. The most continental areas were the earliest to lose the Neanderthals and this happened quite soon. Curiously, the first populations to go were not those from the coldest, northernmost areas but those from cold and *arid* regions. The earliest regional extinction occurred in the Balkan Mountains and the Carpathian Basin between 47 and 46 thousand years ago followed by the Middle East around 45 thousand years ago. In spite of being on the Mediterranean and having had an important Neanderthal population,[7] the Middle East was among the first areas to lose the Neanderthals and I attribute this to the repeated encroachment of desert into the region. Aridity—water shortage—would have been responsible for the Neanderthal extinction in the Middle East.

The negative effect of continental climate is seen clearly in the next two areas to lose Neanderthals—parts of Turkey and northern Europe between 44 and 43 thousand years ago; Turkey is much further south than northern Europe but it is also well east and has a very continental climate. Between 42.5 and 40 thousand years ago, huge areas of Italy, remaining areas of Turkey, the Aegean, Mediterranean France, and inland northern Iberia lost the Neanderthals. The first region of Iberia to be depopulated was the harsh northern interior which has the most continental climatic regime of the Iberian Peninsula.[8] The Neanderthals disappeared from there and the remaining continental areas of the Mediterranean. This means that we should not expect late Neanderthal populations, more recent than 40 thousand years ago, across much of continental Europe or even inland Iberia.[9]

The regions that one would expect to have retained Neanderthals past the threshold of 40 thousand years ago would be parts of Iberia, the coastal areas of the Black Sea, and the Atlantic coast of Europe, and this proves to have been the case. This shows well how coastal areas were the most favourable for the Neanderthals. It would have been in these areas that the effects of continentality and aridity would have been minimized. The Neanderthal populations in north-western Iberia and south-western France were among the last, between 36.5 and 35 thousand years ago, leaving the southern Iberian populations as the last with the south-west being the last stronghold, with Neanderthals surviving there to between 31 and 30 thousand years ago. I then looked at a number of features in each region to see what factors had been the most important to Neanderthal survival.

The three main ones were proximity to the Atlantic Ocean, to a coastline, and to other important regions. Proximity to the Atlantic Ocean was important because it was here that the most humid, least continental climates were found. The importance of this oceanic influence in the south-west of the Iberian Peninsula became clear when we looked at the climate of the whole of the south of the Iberian Peninsula.[10] The present-day climate for the entire area was mapped using GIS.[11] Next we redrew the maps by introducing incremental falls in temperature and rainfall to simulate conditions in the region during a cold and dry glacial. The maps gave us a lot of information. The south-western extreme, which includes Gibraltar, is the mildest and wettest today. The town of Grazalema, just 70 kilometres north of Gorham's Cave in Gibraltar, has the highest annual rainfall of the entire Iberian Peninsula that averages close to 2,000 millimetres. The reason for this phenomenon is that the area is the first to receive the Atlantic storms in between autumn and spring and these discharge their rain on the first high ground that they hit. This water then flows towards the Atlantic Ocean and the Strait of Gibraltar, irrigating drier lowland areas along the way.

What startled us was that, when we simulated a cold and dry period, this south-western corner remained mild and wet, in contrast with higher ground and areas to the east which became veritable cold deserts. When we plotted Neanderthal sites onto these maps we could see how they clustered around the major rivers and the coast and the majority of sites were precisely in the south-west. We realized why the south-western corner of Iberia, with its southerly latitude and proximity to the Atlantic had been the last Neanderthal stronghold: water had been the key factor. This revealing information showed that the Neanderthals had also been limited, like *sapiens* in Africa, by water availability, and why areas well north, such as Atlantic France and Britain, had remained occupied later than continental European, including central and eastern Mediterranean, regions. They were wetter. I will return to this point later in this chapter as it is crucial to understanding the Neanderthals and their survival.

Proximity to coastlines was the second ingredient of the three-part cocktail that I had come up with to explain Neanderthal survival. Coastlines were clearly influential as climate buffers as well as potential sources of fresh water (recall the coastal oases discussed in Chapter 7) and a rich array of foods. I am very conscious of the effect the coast has on microclimate: I live in Gibraltar right on the coast of the former Neanderthal stronghold. Here, summer temperatures, though hot, are much cooler than in the interior and winter temperatures are milder. It never snows in Gibraltar but you can regularly see snow in the mountains nearby. But to see the continental effect I do not need to travel up. Let me give you an example. During the winter of 2012–13, as I was writing this book, I carried out some fieldwork in an inland, low-lying area not far from Gibraltar. In the car I would get there in 40 minutes. A typical January excursion would see me leaving my house an hour before sunrise with the temperature around 12 °C. By the time I got to my study site I had seen frost, my car was warning me of ice on the road, and the temperature would range between −2 °C and +3 °C. By the time I was ready to leave for home, around 4 p.m., the temperature in the study site had risen to 19 °C or 20 °C, similar to what I found on my return home. So, in a single day, the temperature range at home on the coast had been around 8 °C while at my inland study site (at a similar altitude above sea level) the range would have been anything up to 22 °C. In midsummer, my visits to this site have been as brief as possible as the midday temperature always surpassed 40 °C, while it was unusual for it to go beyond 30 °C at home. Life has always been easier on the coast than inland.

The coast also offered additional resources. In 2012 we started excavating Vanguard Cave, another huge cave which is situated about 100 metres north of Gorham's Cave. We targeted this site because it has huge archaeological and palaeontological deposits, around 17 metres thick, and because the Neanderthals lived in this cave too. We are particularly interested because earlier trials have shown that the Neanderthals at Vanguard exploited coastal resources, something which nobody had expected as it had been considered beyond their capacity. We know that at Vanguard Cave the Neanderthals consumed fish, shellfish, monk seals, and at least two species of dolphin[12]—and this may be the tip of the iceberg. Vanguard Cave may offer us some surprises in the future. The use of coastal resources seems to go back as far as the use of such resources by other people in South Africa.[13] When we looked at the Neanderthal site of Bajondillo on the coast by Torremolinos, just 90 kilometres north-east of Gorham's and Vanguard Caves, we found evidence that the Neanderthals there had been consuming marine molluscs around 150 thousand years ago, a date comparable to that claimed to be the earliest human exploitation of marine resources at Pinnacle Point in South Africa (Chapter 7). With a climate buffered from extremes and abundant freshwater and food resources, the coast within the south-western refuge was an important piece of Neanderthal real estate.

The importance of fresh water and marine foods in our evolution has received attention recently.[14] The view that these resources were essential to our development, particularly so as to support our large brains, is probably an overstatement since alternatives to the key nutrients found in these environments are also available in non-aquatic habitats.[15] Living close to fresh water and the coast would have undoubtedly permitted human populations to exploit the foods associated with these environments but that does not mean that they were incapable of surviving and doing well in their absence.

The third element of the Neanderthal survival cocktail was proximity to other Neanderthal strongholds. This would have been important because survival would have been dependent on genetic mixing of populations. Small isolated populations would have stood less chance of long-term success than those which were surrounded by others. The south-west was well placed as it had other important Neanderthal regions of Iberia adjacent to it, inland and along the coast eastwards into the Mediterranean and northwards up the Atlantic coast of Portugal and from there the Cantabrian and south-west French strongholds.

So the Neanderthals survived late in mild and humid places but what was the wider picture of Neanderthal occupation across its Eurasian range? To answer this, my research group looked at Neanderthal sites across the continent and tried to find out what habitats they had occupied in disparate parts of the range.

To do this we looked at the fossil birds which had accumulated alongside the Neanderthals for clues. We found that across this vast range the Neanderthals lived close to sources of fresh water and along coasts where these were available.[16] It seemed that, like the humans of southern Middle Earth, the Neanderthals were tied to water except that such sources were more readily available in Eurasia than in Africa. The oceanic west nearly always provided wet environments and the Neanderthals lived by rivers, streams, and lakes.[17] Further east they would have depended more on snow melt from the high mountains, exploiting the ecological diversity provided by altitude gradients along hillsides, slopes, and valleys.[18] So our earlier results now made sense. As the climates became colder and drier, the most severely affected regions would have been the continental interiors. The Neanderthals would, like the people of southern Middle Earth, have experienced water shortages and it would have been close to the Atlantic and along coasts that water sources would have remained most plentiful and where the Neanderthals would have held on for longest.

Our study of the broad, continental, ecological requirements of the Neanderthals also revealed that they had a close association with rocky habitats,[19] far more than at any other point in our history and it suggests that the predisposition to visit rocky areas and shelter in caves became a part of everyday life in the cool climates of Eurasia after 450 thousand years ago just as it had for people in South Africa. It seems that it was the Neanderthals, equipped with the ability to control fire, who were the first to use caves regularly. When we turned our attention to the vegetation that composed the Neanderthal habitat we found that this typically incorporated trees and open spaces, sometimes but not always bushy thickets. Put together, the Neanderthal lineage behaved in typical *Homo sapiens* fashion, living in areas with trees and open spaces, and with water nearby. They lived close to the coast as well as inland. If there was a distinguishing mark of Neanderthal habitat that set it apart from many other *Homo sapiens* populations, it was their regular use of rocky habitats and caves. Together with their South African contemporaries, the Neanderthals were the first 'cavemen'.

Could we find direct evidence of the impact of water shortages on the Neanderthals? We decided to go back and take another look at Gorham's Cave.[20] After the last Neanderthals, this cave seems to have been abandoned for a while. It seems that a place that had provided optimal living conditions for the Neanderthals for tens of thousands of years suddenly lost its charm. What could have happened? To try to answer this we looked for evidence in other caves around Gibraltar. We also looked at marine cores taken from the seabed for clues of climate behaviour.[21] The results made us jump with excitement.

Wherever we looked in Gibraltar we found that the place had apparently become unliveable precisely when the last Neanderthals left 31.5 thousand years ago; nobody replaced them. Seismic activity was one of two key factors—we found evidence of massive falls of stalactites inside caves, rock collapses that opened up previously sealed caves, and massive landslides and rockfalls. Living inside caves or at the base of cliffs had suddenly become a hazardous business. The second factor was drought. The marine cores picked up evidence of large-scale wind-blown sand transport and of lowered river input into the sea. For a brief moment, the southwest of Iberia became a place to be avoided as three of the four key factors that suited the Neanderthals best—plentiful freshwater supplies, rocky places, and links with other populations—were lost: the oceanic influence temporarily dwindled, caves became dangerous places to visit, and there were no other Neanderthal populations left. All that remained was the coast and it is possible that the adaptability of the Neanderthals allowed them to hang on a little bit longer by exploiting marine and coastal resources when those on land had become hard to find.

Why would the Neanderthals have suffered from drought when the humans in the rest of southern Middle Earth were managing all right in similar conditions? It all has to do with the starting conditions of each lineage and their subsequent evolution in very different ecological settings.[22] The Eurasian Neanderthal lineage split from the southern Middle Earth human lineage ~450 thousand years ago, at around the time when the latter had started to experience severe droughts. As we have seen, in southern Middle Earth people continued to become increasingly light and streamlined in response to the need to move about quickly over large areas, and composite tools, which improved mobility, were developed. The Neanderthal lineage also possessed this technique from a very early age, around 300 thousand years ago.[23] This means that there must have been cultural transmission of information between northern and southern people after the lineage split unless we were to postulate the unlikely scenario that the same technology was independently invented by the two lineages.

The Neanderthal lineage must have experienced similar climatic conditions to other humans in places such as the Middle East—where the two overlapped sometime after 120 thousand years ago—Arabia, and possibly India. Our knowledge of the distribution of populations of *Homo sapiens* in these critical areas is too fragmentary for us to be able to know how far each lineage went into these regions and how much overlap there was. My hunch is that it was in the centre of Middle Earth, east of the Mediterranean and west of the high Asian mountains, that such contact must have taken place in areas where the warm and arid world of the south interacted with the cold and arid world of the north, perhaps in brief moments of improved rainfall.[24] I shall explore this idea further in Chapter 9.

To the north the Neanderthals lived in a very different world, one of trees and open spaces, cliffs and rocky scree slopes, mountains and valleys, rivers and lakes. This covered the northern slopes of the mountains of Middle Earth—in such ranges as the Altai in Siberia, the Zagros (Iran), the Caucasus (between the Black and Caspian Seas), and the Carpathians, Alps, and Pyrenees in Europe—and it was mainly in the Far West, in Atlantic Europe, that such environments spread north and away from the mountains of Middle Earth during warm interglacials. Beyond these areas lay open, flat regions where cold deserts and steppe kept the Neanderthals out. While in southern Middle Earth humans had no choice but to continue adapting biologically and culturally to dry, open plains—they were hemmed in between harsh desert, the sea, and impenetrable jungle—the Neanderthals did have a choice, for a while at least.

Here they pursued an energetic lifestyle of ambush, hunting prey at close quarters with hafted tools in typically *Homo sapiens* environments of trees, open spaces, and water. Their energetic lifestyles promoted the continued development of muscular bodies and strong bones[25] but their hind limb anatomy could no longer cope with efficient endurance running.[26] It was the populations of southern Middle Earth that went down the new route of lightweight bodies with important consequences as we shall see in Chapter 9. The Neanderthals could afford to proceed in this muscular direction. After all, they lived in a world that did not require the far-ranging excursions of their southern cousins. It was this commitment to small-scale territories that set them back thousands of years later. Intelligent humans that they were, it seems that they tried to cope with the rapidly advancing world of treeless cold steppe when it reached them in the west but it was too late. They were overrun except in the extreme south-west where the cold steppe never reached. But their population was by then small and fragmented, and a local water shortage linked with seismic activity finished them off for ever.

9

Global Expansion of the Rain Chasers

70–21 THOUSAND YEARS AGO

According to what is called the 'Out-of-Africa 2' model, during the period from 70 thousand years ago, *Homo sapiens sapiens* spread from Africa and rapidly colonized different parts of the world, including Australia by 50 thousand years ago and Europe by 40 thousand years ago.[1] However the interpretation I present here is somewhat different.

Let us take things one step at a time. What was there before and what was there after the expansion? I will not go into the timing of the geographical expansion for now at least because it would only complicate the picture. Before the spreading-out of a population of *Homo sapiens* across many parts of the Old World we had a lineage that occupied parts of southern Middle Earth. That lineage, which may well have had its continental variants, had become the lightly built, highly mobile rain chasers (Chapter 7). So far we have taken for granted that this lineage—the rain chasers—originated and evolved in Africa. We should not ignore the important fact that the climatic conditions that affected parts of northern and eastern Africa, and which we have seen were critical in the evolution of humans, also affected Arabia, the Middle East, and India in similar fashion. In fact, the conditions in Arabia would have been closer to north-east Africa than the latter would have been to South Africa. Yet our Africa-focused mindset seems to prevent us from comparing similar zones if they lie outside the geographical limits of the continent.[2]

The absence of fossils in Arabia and India has forced the 'African viewpoint'. In any case the number of African fossils is very small and largely from the rich deposits in the north-east, particularly Ethiopia. Besides falling under similar climatic regimes, these sites are next door to the Arabian Peninsula and the Middle East and they are even closer to Iraq, Iran, and India than they are to southern South Africa. It would make sense, simply because of proximity, to expect that north-east African, Middle Eastern, Arabian, and even Indian populations interchanged genes and ideas on a regular basis, more so than they would have done with distant parts of Africa, such as the south or the Atlantic north-west (present-day Morocco). It may sound surprising but central Asia is also closer to north-east Africa than the latter is to South Africa. But the primary barrier to interchange with southern Africa would, in this case, be climate as changes in latitude would mean important shifts in temperature regimes. When conditions were favourable, during warm and wet periods, interchange between central Asia, Iran, and Iraq should have been possible. These could have been routes of cultural interchange, a way perhaps in which composite tools reached the Neanderthals and the rain chasers; from whom to whom and from where to where we just do not know.

As the northern part of Middle Earth came increasingly under the influence of the glaciations, its active part—where the cauldron of human activity was located—was reduced to the vast area occupied today by the Sahara, east to Ethiopia and the Middle East, then east again across Arabia, Iraq, and Iran to the Indian subcontinent. This was the southern Middle Earth that I have been referring to so far. Human populations would have readily exchanged genes and culture within this area which, during humid periods, would have been an Eden of lakes and rivers.[3] During dry periods, human populations would have become isolated from each other in wetland refuges; if the isolation was prolonged then each population would develop its own biological and cultural identity. When populations met once more, these biological and cultural traits would have been interchanged, hence the mosaic nature of our species that was evident in Chapter 7. As this process of expansion and contraction was repeated across southern Middle Earth, genetic and cultural signals would merge or be superimposed over others. Such a pattern explains why there is so much discussion, and confusion, regarding when 'modern humans' left Africa, how they did it, and which routes they followed.[4] There simply was no one-off Out-of-Africa; people had occupied the entire southern Middle Earth since the earliest times (Chapter 5) and different populations across this region had had contact, on and off, for the better part of 1.8 million years. The changes in the geographical range of humans did not, in this view, start with the build-up to the Last Glacial Maximum around 100 thousand years ago either, as Stewart and Stringer have suggested;[5] they were not movements in the way human migrations are envisaged by many current authors[6] and they did not have a specific commencement date either.

Where the flow of genes and ideas between the rain chasers and their neighbours broke down and only opened sporadically, was in south-east Asia, where rainforests were a major barrier and between Iraq, Iran, and central Asia, where cold and dry conditions would have limited the opportunities for exchange. These did happen, the genes we have acquired from Neanderthals and the Denisovans of Siberia[7] bear testimony to at least two such occasions. Further west, the Mediterranean and its mountains were probably a permanent barrier, as were the Himalayas and associated mountain ranges further east. We should see southern Middle Earth as the region where a human lineage was fine-tuned into an endurance runner and long-distance walker. It was also a pump (or source region, see Chapter 6) that retained a permanent human presence throughout, albeit patchily at times, that occasionally fed areas beyond its borders. The areas beyond its borders were southern Africa, south-east Asia, and central Asia.

South-east Asia was a second pump that fed areas to the north, particularly China, when rainforests opened up and cold environments shifted northwards; during glaciations this region acted as refuge for human populations. South-east Asia may have been the original source of *Homo sapiens*, given the early Javan dates that are comparable in age to the earliest African ones, or it was colonized very early instead. I suspect the latter. Once established, the population remained for at least 1.7 million years, largely isolated from events in the rest of southern Middle Earth. Any subsequent contact between these populations remains obscure. We only find evidence of the rain chasers in south-east Asia around 45 thousand years ago[8] but the arrival of these people in uninhabited Australia between 50 and 46 thousand years ago[9] indicates that they had reached south-east Asia before this time.

We just do not know how many more times human populations reached south-east Asia after the first wave around 1.8 million years ago but given the many times that savannah corridors would have opened up in the area as climate cooled and became drier, it seems unlikely that we have been able to detect the only expansions that ever happened.

Let us go back to the Denisovans. They shared an ancestry with the Neanderthals and their remains have been found in Siberia, which has led to the interpretation that they somehow originated or had their core area in northern Eurasia, east of the Neanderthals. But analysis of their genetic make-up reveals that they had a lot in common with present-day Melanesians and little with Europeans and Chinese people. This has led to suggestions that their geographical range shifted southwards during cold and dry climatic pulses. What if it wasn't like this? What if the common ancestor of Neanderthals and Denisovans lived somewhere in southern Middle Earth? From there the Neanderthal lineage could have entered central Asia and spread afterwards into Europe and the Denisovan lineage would have penetrated into south-east Asia and then northwards into Siberia, via a route east of the Himalayas.

This pattern of geographical spread would fit well with our model of sources and sinks. If south-east Asia was, as we proposed back in 2000,[10] a major source of human expansions then a Denisovan origin in south-east Asia and subsequent expansion northwards should not surprise us. This would not have been unusual either. When I looked at the pattern of distribution of Palaearctic birds[11] I found that many diverse south-east Asian groups had contributed populations and species northwards. Some of these species had initially adapted from tropical to more temperate climates by moving up the slopes of the Himalayas.[12] When cooler conditions shifted climate belts southwards there would have been moments when comparable altitudinal and latitudinal climatic belts would have merged, releasing the populations trapped in a particular altitude zone. Then, when the climate belts shifted north once again, they would have carried new species with them. Once established on the Eurasian plains many of these species (including such diverse groups as crows, bullfinches, and orioles), well adapted to the climatic conditions, rapidly spread westwards across vast distances that had few topographical barriers, many reaching all the way to western Europe. I called this the Himalayan launchpad and found many examples among bird species (Fig. 5).

The point that is important here is that these populations spread north during climate warming, but tracking temperate, not tropical, belts that they had adapted to on the slopes of the Himalayas.[13] Once in northern Eurasia they did not have to adapt to new conditions—they were already used to them and had tracked them north. The beauty of this model is that it allows populations time to adapt to new conditions within stable source areas. We saw in Chapter 6 how populations of early humans in tropical south-east Asia had moved up mountains in search of water when conditions became drier in the lowlands so we have a good basis for our model and it is, once more, linked directly with water. So I propose that the Denisovans were one such population, one that spread northwards and then westwards following temperate belts, and it is no coincidence that we discover them on the slopes of the Altai Mountains, on the other side of the Himalayas, and associated mountain ranges.

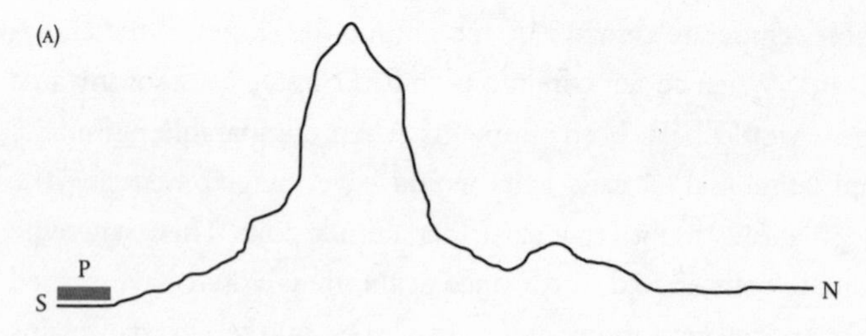

FIGURE 5A. Sequence showing how the Himalayan launchpad works: 5a and 5b show an intitial, tropical population; 5C and 5D show the population having dispersed up mountain slopes, during a cool and arid phase, in search of humid locations. Comparable climatic environments are available, but are disconnected and empty, to the north; 5E and 5F show a cooling that brings the climatic belt and humans down the hills. The belt to the north spreads south and merges with the altitude band allowing humans to spread; 5G and 5H show a subsequent warming with the belts once again separated from each other. This time humans occupy the altitude and latitude belts. Once in the latitude belt, human populations (P) can spread horizontally across large areas of land while those in the south are confined to their mountain refuges.

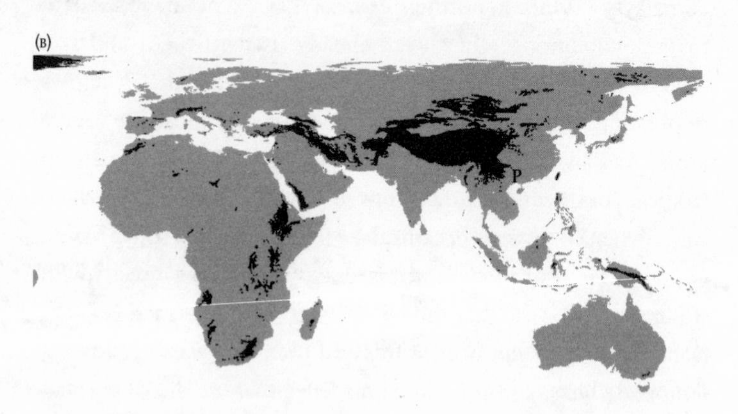

FIGURE 5B. Continued

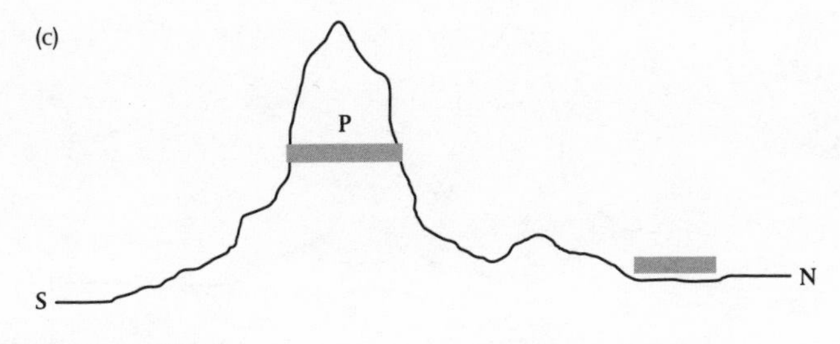

FIGURE 5C. Continued

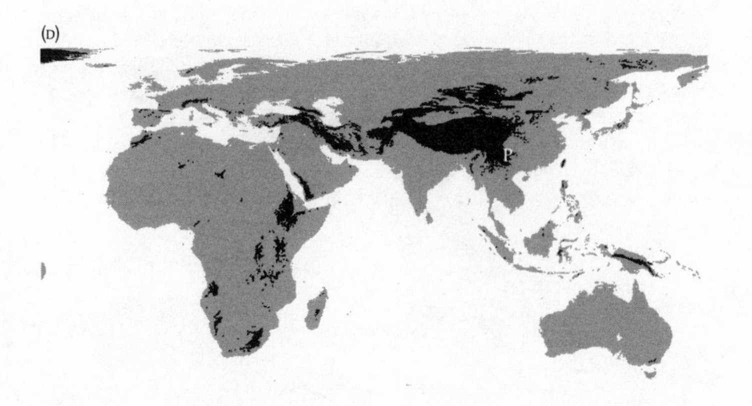

FIGURE 5D. Continued

The Densiovans contributed genes to present-day Melanesians but not to Chinese or Australians. How can we explain this disparity? If the Denisovans entered south-east Asia from southern Middle Earth during a cold and dry phase when the rainforest opened up, we would expect them to have colonized mountain slopes where they could find water. Later arrivals of rain chasers probably behaved in a similar manner allowing opportunities for genetic interchange. But we have also seen how earlier human populations had also skirted round the expanding rainforest

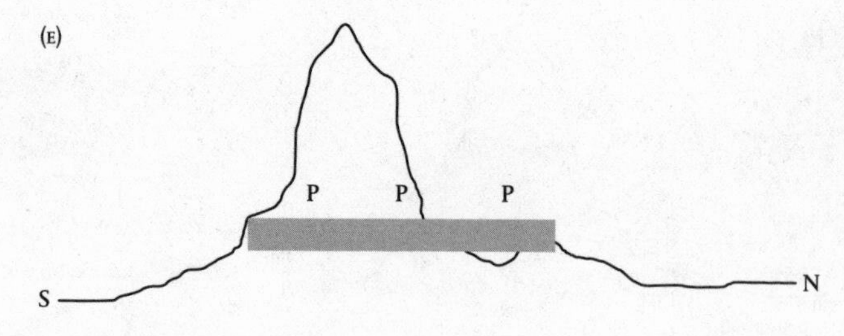

FIGURE 5E. Continued

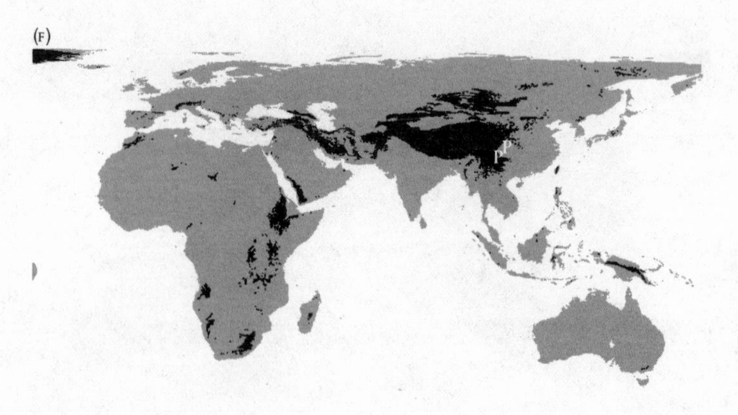

FIGURE 5F. Continued

during warm and wet conditions by concentrating on the coast. We can see how mountain slope and coastal populations would have become segregated during such times. So a population of coastal humans may have proceeded towards Australia having had no contact with the mountain Denisovans. And, if the Denisovans followed a specific climatic belt, this could have kept them apart from later waves of people that entered China.

What of the Neanderthal lineage—could they have entered Eurasia via a similar mountain launchpad route? I think so, except

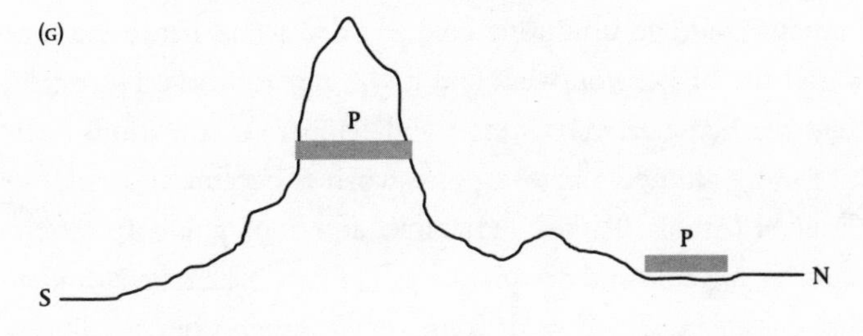

FIGURE 5G. Continued

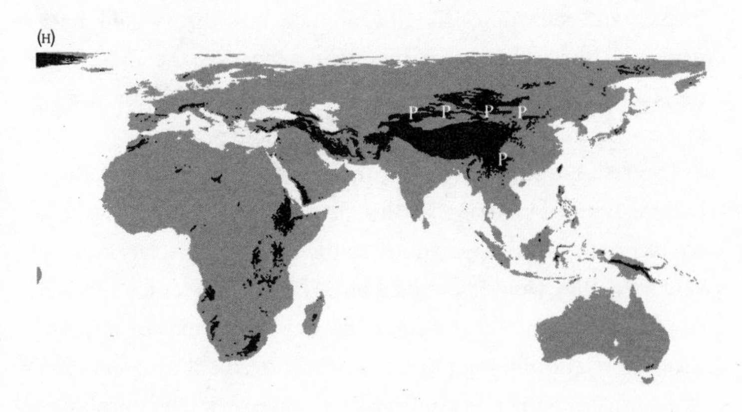

FIGURE 5H. Continued

that they would have accessed Eurasia west, not east, of the Himalayas and the opportunities would have been limited in comparison. When they met the Denisovan lineage they would have been occupying a similar climatic belt and would have come into direct contact with each other; and they were similar enough still to have been able to exchange genes. Let us explore this a little bit further.

The Himalayan launchpad relies on the warm tropical conditions of its southern slopes. Here the Denisovans and other humans could go up during cold and dry events because it was wetter the higher you went and it was not too cold. Essentially they reached relatively humid and temperate conditions with increasing altitude, up to a point when it became too cold. In the west (Arabia, Turkey, Syria, Iraq, and Iran), going up mountains during cold and dry phases would have taken the Neanderthal lineage people towards wetter environments but conditions would have been much colder than in the south-east Asian tropics and subtropics. So the altitude belt that would have permitted survival in such mountain ranges as the Zagros or the Caucasus would have been much lower down, narrower, and restricted.

The result would have been that at many times populations became trapped in particular altitude zones and, if there was little contact with other populations or the periods of isolation were very long, they would have died out. The ideal places for survival would have been where mountains and valleys existed in close succession, affording opportunities for movement up and down with climate change. Turkey, with its high plateau, was probably unsuitable.[14] The populations inhabiting mountain refuges in the Arabian Mountains[15] to the south would have been too detached but the Caucasus[16] and Zagros[17] Mountains would appear to have some survival potential within limited altitude belts. The nature of these locations would have promoted regional cultures, evident in the boundaries between stone-tool technologies in the Caucasus.[18] It is not surprising either, given the high risk of population isolation and the Neanderthal abhorrence of extremely continental climates, that they seem to have died out at an early stage in these mountains.[19]

So the Neanderthal lineage would have penetrated into central Asia in the first instance, probably via present-day Iran or across the Caucasus, by moving up and down altitude zones as climate alternated between cold and dry (moving up the slopes in search of water but limited by cold) and warm and wet (moving down slopes into inter-montane valleys and plateaux). The cold and dry periods would be spent in wet hillside refuges that were probably isolated from each other and the warm wet periods would have launched the populations northwards. Once in the northern mountains, such as the Siberian Altai, they would have kept to particular climate belts which would, again, have moved up and down with the vicissitudes of climate. Penetration of the cold steppes and deserts would have been extremely difficult for these populations.[20]

We do not know how many times the Himalayas and Caucasus–Zagros might have worked as launchpads. It probably happened frequently and with variable success. We will probably never discover the failures, only the successes: we have known of the Neanderthals for a long time but we have only just discovered the Denisovans. We have little information about these people who were the contemporaries of the Neanderthals. We know of their existence from analysis of their DNA from fragmentary and undiagnostic bones recovered from Denisova Cave in Siberia. They must have once had a widespread geographical range and their genetic signal persists today in a number of human populations, particularly Polynesians. There must have been others. One successful entry seems to have happened relatively recently, bringing a population of *sapiens* into central Asia around 45 thousand years ago.[21] This entry happened during one of the coldest and driest periods of the Earth's recent history, an apparent paradox.

There is a simple possible explanation. As would have happened many times before, a population of rain chasers living south of the Caucasus–Zagros launchpad was hit hard by arid conditions. Each time this happened the resultant move up into the hills for water would have been magnified as some human populations became better at coping with water shortages. The biological and cultural improvements for dealing with water shortages would have allowed humans to survive, as the Neanderthal ancestors had previously done, in higher altitude belts. The climate was colder and drier than ever before but humans would have been better prepared than their predecessors, so much so that these rain chasers were even able to spread into the drier interiors and eventually reach central Asia.

Once in central Asia, the behaviour of the new human population arriving from southern Middle Earth is a classic example of the kind of ecological release that I observed among many species of birds. Once in the new latitude band, movement west and east would have been very fast, taking these humans to the coasts of the Atlantic and Pacific in a relatively brief period. The plains-adapted humans found a rich world of herds of herbivores, one that could be accessed readily by a highly mobile hunter with a correspondingly lightweight toolkit which, by 45 thousand years ago, included projectile technology as well as composite tools.[22] Finding water may still have been a problem once in these high latitudes but now they had new opportunities and solutions. When not readily available, ice and fire would have been combined to generate the vital liquid. The adaptation for moving around over vast distances in search of water, the defining feature of the rain chasers, was now being put to good use in tracking down herds of reindeer and other migratory herbivores. Their ability to do what the bulky Neanderthals could not do so well was an outcome of having had a rain-chasing career. Importantly, it allowed them to break away from the restrictive altitude belts of the high mountains that ran from Iberia to China and onto the plains beyond.

In the process, they would have come across Neanderthals but not Denisovans. We know this from the genetic contribution of the Neanderthals to the European human genome and the absence of such a contribution from the Denisovans. This is a strong indication that the entry was from the Caucasus–Zagros launchpad into southern Siberia. Somewhere in between, before the entry into Europe or the uninhabited plains, Neanderthals and rain chasers exchanged genes. Once on the plains, the rain chasers were on their own, at least until they came into contact with populations of Neanderthals in the west where the cold mammoth-steppe was reaching the latter's receding world. It was only in the oceanic west that the Neanderthals had penetrated the plains to the north of Middle Earth and as these Atlantic forests were becoming open steppe, the Neanderthals retreated and the newly arrived humans advanced. The Neanderthals survived in refuges, deep in the warmest and wettest part of Middle Earth which also happened to be the last places that the newcomers reached.

When we looked at the habitat of the Neanderthals using birds as indicators (see Chapter 8), we found the consistent trend that characterized *Homo sapiens* throughout its career: it lived in places with water, trees, and open spaces in combination. We also found that the Neanderthals were the first to exploit rocky habitats to the full. The pattern of substituting rocky places for trees as habitats is another feature that I identified in the success of a large number of bird species in the Palaearctic.[23] Many species belonging to families of tree nesting birds had taken to nesting on rocky surfaces.[24] The adaptation catapulted their east–west spread along the mountain chains of Middle Earth and, in some cases, south along the Rift Valley. The Neanderthal lineage behaved in similar fashion once it took to the world of rocks and caves, a world which would have characterized the mountain slopes that had marked their point of entry into Eurasia and their subsequent expansion with the continent.

We then looked at bird indicators of the habitat of the rain chasers in Eurasia. We found many similarities with the Neanderthals but also some very telling differences. The habitat specificity of the Neanderthals, so typical of all *Homo sapiens*, fizzled out when looking at the newcomers in Eurasia. Places with trees and open spaces were included but so were places without trees. Water was never forgotten and rocky habitats were also included but the ability to cut across vast expanses of openness was a new feature that characterized the rain chasers in Eurasia, just as it probably did where they originated, somewhere in the core of southern Middle Earth. It was made possible by 2 million years of evolution of a lightweight body built for endurance running and the development of portable projectile technology.

At the time that these events were happening in central Asia, between 45 and 40 thousand years ago, humans had been in Australia for at least 15–20 thousand years. That they should have taken so much longer to disperse from southern Middle

Earth into nearby Eurasia than to distant Australia seems odd. My own view is that entering central Asia took these humans into a world of other humans, Neanderthals and Denisovans at least, who were on familiar home ground. If we take the view, which I advocate, that all these peoples were part of the large widespread *Homo sapiens* family, then we would not expect the locals to have allowed themselves to be overrun by the newcomers. That scenario would only occur if the immigrants had some kind of superiority over the residents. Now, if we take the view that the rain chasers were more intelligent than the Neanderthals, rapid takeover and domination might be expected, but that does not then explain why they should have taken so long to penetrate into Neanderthal territory. My own view is that the Neanderthals were doing very well on the slopes of the Zagros, Caucasus, Altai, and other ranges and they simply kept the newcomers at bay. This need not have been an active defence of territory, although some of that would have surely happened, but a simple process of preventing their demographic expansion by limiting their access to resources. Clearly they met along the way and interchanged genes in the process, as we have seen.

Those entering Australia had no such difficulty. Australia was uninhabited and, once in, there was no resistance of this kind. There were people along the way, in south-east Asia, where early human populations would have been around from the very beginning. Why did this particular expanding population of humans not meet resistance from the locals? I have already suggested how they would have, unconsciously, avoided the Denisovans who would have been occupying humid temperate belts in south-east Asia during cold and dry conditions. These would have been the conditions that would have opened up the rainforest and permitted an expansion of the rain chasers into south-east Asia. My hunch is that they may never have come across anyone at all or, at most, a small and highly scattered population that could not have had the impact on the immigrants that the Neanderthals did in Eurasia. The reason for that has to be the impact of the Toba Volcano eruption.[25]

The Toba super-eruption in Sumatra of 74 thousand years ago would have had a devastating effect on populations that were living in the south-east Asian heartland. The event was one of the world's largest known eruptions and is thought to have generated a volcanic winter that may have lasted up to 10 years, causing also a 1,000-year-long global climate-cooling. In a recent paper Mike Petraglia at Oxford has downplayed the impact of Toba on people living in India and further away from Toba.[26] While it makes sense that at least some populations away from the eruption may have survived, the situation closer to home would have been far different especially when the post-eruption effects, notably climate cooling and aridification, are considered.[27] The consequent opening up of the rainforest let the rain chasers in. Their adaptability to arid environments would have enabled them to survive the super-eruption and subsequent aridity in southern Middle Earth and stage an early recovery. When the gates of south-east Asia opened up, humans were ready to use this new launchpad[28] to catapult themselves in.

We have little evidence of the rain chasers in south-east Asia, other than at Niah Cave on the island of Borneo. Here people exploited a variety of environments including, unusually, the rainforest itself; but they seem to have done so from the river networks and there is evidence that suggests that these people started to open up the landscape with the use of fire. This would be the first case in our history of habitat management, curiously and significantly turning dense forest into places that combined the ancient, tried-and-tested formula of water/trees/open-spaces. The human presence at Niah Cave would be quite late at 45 thousand years ago but there is increasing evidence that humans may have been in the region by 67 thousand years ago,[29] after the super-eruption.

The nature of the geography, climate, and vegetation of south-east Asia is such that it would have promoted life along rivers and coasts during warm and wet phases when rainforest engulfed the land. We saw this happening to humans in the early stages of our evolution, with the first recorded sea-crossing to Flores. It is probably no coincidence that a similar way of responding to such conditions by the new populations got them to new worlds. The arrival of humans in Australia was probably the consequence of a population expansion along rivers and coasts by people used to living in arid lands, and its ultimate cause was a giant volcanic explosion.

The thrust of this book so far has been that there has been a long interrelationship between climate change and human evolution and that the main driver behind this story has been water. The world has been getting progressively more arid and a lineage that led to all of us in the world today managed to survive by constantly adapting to this changing world. These ideas can be drawn together in what I have called the Water Optimization Hypothesis.[30]

This hypothesis proposes that *Homo sapiens* selected environments that fitted into climates that fell within an intermediate range of rainfall regimes, between the hyper-humid rainforests at one end and the hyper-arid deserts at the other (Fig. 6). The closed forests would have been unsuitable because access to resources

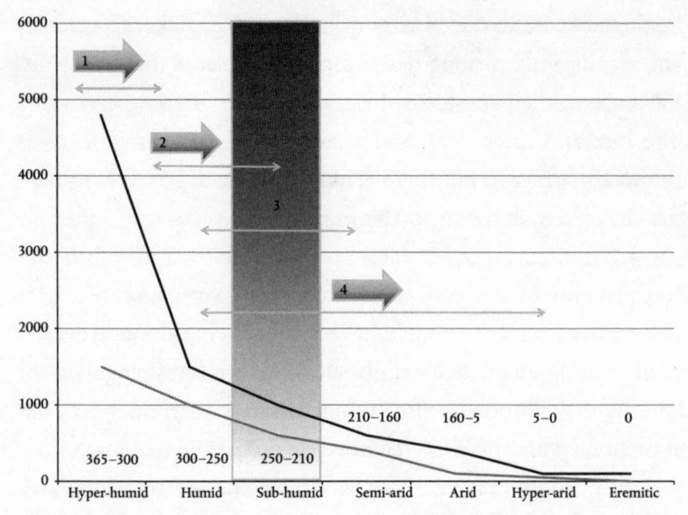

FIGURE 6. The Water Optimization Model. The x-axis shows the range of humidity regimes available to humans. The y-axis shows annual rainfall (mm) with the black line showing the maximum and the grey line the minimum annual rainfall for each humidity category. The numbers above each category indicate the typical number of days of rain/year in that category. Arrow 1 represents the climatic shift of the earliest hominids around 7–6 million years ago with the double arrow indicating the range of climates tolerated; arrow 2 represents a later trend corresponding to the australopithecines; the bar numbered 3 indicates the optimal human habitat, already observable at 1.8 million years ago, with the double arrow indicating the range of tolerance; arrow 4 shows the progressive shift of humans after 1.8 million years ago towards increasingly arid climates while keeping sub-humid as optimal and amplifying their tolerance (double arrows) in the process.

would have been very difficult and the most arid deserts would have been just as bad because of lack of water and other resources. The middle ground would have provided ample opportunities for humans to stick to their favourite formula of trees/open-spaces/water. The optimal conditions would have been provided by sub-humid regimes which offered rainfall between 600 and 1,000 millimetres/year with 210 to 250 rain days/year. These regions would have offered seasonal rainfall, in some places in the summer and in others in the winter.

Climates at either end of sub-humid would have offered suitable places in decreasing importance the further away that we moved from the optimum. There would have been fewer opportunities in the direction of the humid end of the gradient than towards the arid end, because populations living within the humid regime[31] would have soon encountered habitats, such as dense woodland and forest, which were difficult places in which to hunt and gather successfully. In other words their favourite formula of trees/open-spaces/water would have soon been lost. In the other direction, the situation would have been better as populations would have experienced a range of options, within the preferred trees/open-spaces/water combination, in semi-arid and arid climates.[32] And this was the direction of physical and behavioural adaptation that, as we have seen, marked the evolution of *Homo sapiens*. The irony is that, in spite of their dependence on water, the climates that offered the highest amounts of rain were actually worse for humans than the drier regimes. Tethered to water but requiring areas with trees and open spaces, humans were constantly being pushed in the direction of sub-humid to arid climates which in turn provided the pressure to develop means of dealing with highly dispersed, seasonal, and ephemeral water sources. That is how *Homo sapiens* evolved.

By about 200 thousand years ago, we had come a long way since the early days of the fruit-eating primate of the rainforest. What did the last stages of this evolution look like? What key features characterized the geographical distribution of *Homo sapiens* from around 200 thousand years ago (when people claim to observe fossils that look like us) and 10 thousand years ago (the start of present climate and the origins of agriculture)? In order to find the key attributes of these hunter-gatherer groups, I looked for the very general features that characterized the environments where *Homo sapiens* had lived. For this purpose, it does not matter very much, for example, if a site was close to a river or a lake; what matters is that the site was near a source of fresh water. Using the same logic, it does not matter in terms of diet if the fauna associated with a particular site had deer or antelope as both represented similarly sized packages of meat. It can matter if the different species represent different kinds of habitat but I was not looking at that aspect in this particular study. In the sense of Claude Lévi-Strauss,[33] I was looking for 'universals', the big picture features that characterized all of our ancestors.

To piece this together, it was necessary to trawl the literature for sites that contained good descriptions of the environments surrounding sites occupied by *Homo sapiens*. These had to be selected according to strict criteria. The main ones were that the site showed evidence of repeated occupation—it had a series of archaeological layers that told us that people had visited the site frequently—or evidence that the site was large and covered an important area of ground. The latter usually applied to open air sites, as caves would have been naturally confined. The reason for these selection criteria takes me back to sources and sinks (Chapter 6). A population is said to be a source population if it produces an annual surplus of individuals that emigrate from the natal population. A source should therefore be a good site if its conditions are such that it generates a population surplus. Then there are sinks. These are maintained by a supply of immigrants from source areas; a sink population is in demographic deficit because conditions are barely suitable to maintain existing individuals.

To understand the dynamics of the population it is necessary to distinguish between sources and sinks. A human presence at an archaeological site need not represent, as is often interpreted, that the place was a good one. What if it was just kept propped up by immigrants? Such populations need to be distinguished from source ones—a tall order when looking at archaeological sites that could be tens of thousands of years old. So I chose the two criteria that I have just described as a measure of a good site worthy of a source population.[34] Surely if a place was visited repeatedly or it occupied a large area—suggesting a large population—it stood a good chance of being a source? This could be wrong but it is the best we can do.

The first observation was that the sites were not randomly distributed across the globe but were, instead, concentrated in hot spots.[35] Every one of the 357 sites turned out to have been close to fresh water or the coast.[36] Humans sought out sites close to fresh water wherever they were in the world. Neanderthals and the southern newcomers proved to be no different from each other in this respect. There could be no doubt that the geographical distribution of humans had been tethered to sources of fresh water and it followed that global expansion and survival in refuges had to have followed such water sources.

Were there regional differences between sites? A southern group of sites emerged which was characterized by a mix of cave and open-air sites, a high representation of coastal sites and resources, including large marine mammals, and a high number of inland freshwater sites with large terrestrial game animals. One surprise was the similarity between south-western Iberian sites and those in

Africa and Australia rather than with sites in the rest of Europe. The reason would seem to be that south-western Iberia has a highly seasonal and annually variable rainfall regime, situated within the arid to sub-humid climatic range in warm climates.[37] These characteristics would have been shared for long periods across the southern flanks of Middle Earth and, during warmer, earlier periods of the Pleistocene and Pliocene, also along its northern flank.

These climatic regimes were precisely those that would have provided optimal human living conditions. Regions with these climatic characteristics were the crucibles of human evolution: vast areas of the Sahara, the Arabian, and Indian Deserts of today[38] would have been important areas within the optimal life zones of Pleistocene humans. Further east, the connection with Australia would have been sporadic depending on times when the rainforest barriers opened up during dry climatic phases.[39] Once colonized by humans, biological and cultural changes in Australia would have proceeded in relative isolation. In the Late Pleistocene, south-eastern Australia, southern Africa, and south-western Iberia would have represented the extremes of this vast world. The New World was only just in the process of colonization, largely via the coasts, and so would have not yet been part of this scheme.

A northern group of sites, representing an occupation after 45 thousand years ago, contrasted with the southern group and was dominated by open air sites over cave sites. There were very few coastal sites but, instead, there was a high representation of inland freshwater sites with terrestrial mammals. These sites were spread across the plains of northern Eurasia and North America and were representative of the geographical expansion of *Homo sapiens* across northern Eurasia and into North America. These differences between regions—other regions were identified[40]—revealed the adaptability of *Homo sapiens* to different conditions. In some areas they concentrated activities on the coast and exploited its resources; in others they lived in caves and took a wide range of terrestrial animals including those that lived in the rocky habitats close by; yet in others they lived in open air sites on vast plains where large herds of terrestrial mammals were the main target. But for all this adaptability, the common denominator of all sites, the one that really dictated where to live successfully, was water.

The Water Optimization Model predicts that the continuing adaptation of humans to an increasingly arid world permitted them to exploit semi-arid and arid areas. Here their skills at finding water were tested like never before. In Chapter 10 I will explore how they survived.

10

Australia

Australia is the world's driest inhabited continent. It offers us a unique opportunity to observe how humans took on the challenge of living in an arid world and succeeding in it. This was one of humanity's greatest achievements and yet it is never featured as such in the histories of the world. The lack of understanding of the way in which indigenous peoples succeeded in making a living in some of the toughest environments on Earth goes back to the earliest days of colonizers and evangelists who saw backward, uncivilized peoples when they should have been in awe at the way in which these folk had managed to survive in situations in which the colonizers simply would not have stood a chance, even with their technology. Worst still, the colonizers insisted on changing the indigenous people, ruining tens of thousands of years of evolution and adaptation in a matter of decades. In Australia, the civilized colonizers were the most terrible environmental perturbation ever to have descended upon the native people of that continent. It is particularly sad and ironic that people who had lived through some of the most severe environmental challenges that any humans have ever had to endure succumbed in the end to others of their own kind who chose to see themselves as different and superior.

The first people to arrive in Australia must have crossed a significant stretch of water, by boat or raft, at least 50 thousand years ago, if not earlier.[1] These dates are in accord with the spread into south-east Asia from India after the Toba eruption (Chapter 9). They came from south-east Asia but their origin was somewhere in the core of Middle Earth. We know little else about them other than that they must have been quite used to living along the coast and that they brought the knowledge of controlling fire with them. At Niah Cave in Borneo they would use that knowledge precisely to burn down the encroaching rainforest,[2] converting a closed environment into the preferred one of trees and open spaces, with fresh water close at hand in the extensive river network around the cave. These skills in adapting their environment were already with the people that reached Australia. Australia, though, was a much greater challenge.

Not surprisingly the Australian coast seems to have attracted early settlers. Having come across the sea from similar coasts to the north or north-west, the tropical shores of northern Australia would have seemed familiar and the immigrants would have known how to live there, including how and where to find fresh water. The expansion into the interior of the continent has been the subject of much debate:[3] one view proposes a rapid expansion inland, a second prefers an early spread along coasts and a gradual entry inland following the major river systems, and a third proposes an entry into the interior via well-watered regions, penetrating all areas except the arid core of the continent, which was reached much later. The discovery of human occupation of the arid Central Desert at Puritjarra, west of Alice Springs, perhaps going as far back as 35 thousand years,[4] does indicate that while the early dispersal of humans in Australia probably followed coasts and major river systems, the interior was reached at a relatively early stage.[5] The Out-of-Africa coastal dispersal hypothesis, by which human populations spread across the world and into Australia following the coasts, may therefore not be the whole story.[6] In the case of Australia, we should remind ourselves that the people arriving there had an almost 2-million-year-old heritage of adaptation to an increasingly arid world and were well versed in the art of finding water over large distances.

The most extreme and challenging conditions that humans would have faced would have been in the arid zone of Central Australia at the height of the last Ice Age (the Last Glacial Maximum, LGM), around 21/22 thousand years ago.[7] Many locations in the region, such as the Kulpi Mara and Serpent's Glen rock shelters, were abandoned at this time.[8] These sites were situated within the freshwater riverine core of the Central Australian ranges and its well-watered hinterland. It seems that people entered these ranges at a time when there would have been more abundant resources, including large reservoirs of surface water and aquatic animals, than during the LGM. Even today they provide reliable potable water. The ranges stood out above the surrounding dune fields and sand plains and could be seen for distances of tens of kilometres.

Not all sites within the arid zone were abandoned during the LGM. The rock shelter at Puritjarra was occupied throughout.[9] This rock shelter is in a strategic position as it is located near the only permanent water source in the Cleland Hills. It is part of a sandstone escarpment and faces out onto shrubland and woodland, giving access to important resource areas as well as a range of micro-habitats along the edge of the escarpment itself. Puritjarra clearly fitted the trees/open-spaces/water specification of *Homo sapiens.* The archaeology at this rock shelter tells us that during the LGM, unlike at Kulpi Mara and Serpent's Glen, small groups of highly mobile people continued to visit the site in spite of the arid conditions, though sporadically and for brief periods only. The people of Puritjarra show us that the arid desert was not abandoned altogether, a testimony to the achievement of humans as desert-dwellers. If it could be achieved in Australia, clearly it could have been achieved in the Saharan, Arabian, and Indian deserts which were fully developed by then.

The western part of Central Australia, where Puritjarra is situated, is diverse with sand plains, dune fields, stony desert, salt lakes, and rock outcrops. Scattered across this landscape is an archipelago of small rocky ranges, each one always visible on the horizon from the previous one.[10] The landscape between the hills was dominated by open vegetation, with scattered trees and shrubs, even at the height of the LGM. Provided water could be found, as in Puritjarra, humans would have been able to make a living. They seem to have been successful by having large territories and moving around them in small, highly mobile groups of perhaps 10 to 20 people. They were capable of adjusting to a wide range of circumstances within the arid zone. Their opportunistic and flexible responses to changing conditions, linked to an intimate knowledge of the terrain, allowed them to quickly focus on local resources. They could use this experience also when entering new terrain. With these skills they were able to maintain a low-density population throughout the worst conditions; low density seems to have been regulated by the distribution and spacing of permanent water, the productivity of the country which became accessible from these water sources, and the distribution and persistence of seasonal water.

Australia offers us a completely different perspective in our quest to see how people dealt with extremely arid environments. This new window, which complements the archaeological one, is much more detailed because it gives a direct view of people living in the arid interior of Australia. Because contact between some indigenous people and western settlers happened relatively late in some places, we have first-hand accounts of how the Aboriginal people lived and survived in places too hostile for westerners. We always have to be careful in interpreting such observations especially as the influence of early contact cannot be readily dismissed[11] and we cannot fall into the trap, either, of seeing the behaviour of desert inhabitants as static and unchanging through time.[12] It does not follow that the way desert people behaved a few centuries ago was exactly the same as it would have been thousands of years ago but what we are able to take home is first-hand information on how people were able to survive in such environments. That was the crowning achievement[13] of almost 2 million years of evolution and the colonizers failed to recognize it as such when they came across it. It is small consolation that some enlightened people at least recorded what they saw. One such account, which I will draw from in the lines that follow, is that of Professor Robert Tonkinson,[14] who met surviving groups of nomadic Mardu people in the sixties.

How did the Mardu survive in the great Western Desert of Australia? Not surprisingly, the desert adaptation of the Mardu was dependent on living in small groups that were scattered across the landscape. The system kept a low density of population, which was the key to the sustainability of the Mardu economy. Water was at the heart of their world: the Mardu would move in response to rain, which they could detect across vast distances using visual cues, and exploit food sources, such as fish, waterbirds, and fresh plant growth, which came with it. The groups, their size changing according to local conditions and food availability, were very flexible and could move about their territory at short notice. This quick-response tactic allowed the Mardu to exploit different areas rapidly as they became available during the course of the year or even over a period of years.

The flexible system of the Mardu, being able to adjust group size and location as the need arose, was not enough in itself to ensure survival. The Mardu had an exhaustive and very detailed knowledge of the environment and its resources and they knew well how to procure and exploit these. They were first-class naturalists; after all their lives depended on it. The amassing of the collective knowledge was transmitted to the next generation at an early stage. Children developed a breathtaking ability to read correctly the variety of signals that the surroundings produced, for example animal tracks. The long-term importance of this inbuilt database was in time-saving when out hunting or gathering, critical in times of shortage.

What did the Mardu hunt and gather and how did they go about it? Contrary to expectation, the Mardu were not occupied in finding food in the desert every minute of daylight. Their intimate knowledge of their surroundings allowed them much free time, food-collecting activities rarely occupying them more than half the day in the worst cases. Hunting activities were also leisurely, with little sense of urgency. Decisions regarding what to do were taken each day, by men and women collectively, and depended on weather conditions, how much food was left over from the previous day in the camp, and how far available food sources were located. There was also an element of individual preference.

Men and youths sought large game, on their own or in pairs, while women and children collected plant foods and small animals, such as lizards. The women went out in groups, sharing childminding as they undertook their daily activities; it was these collecting activities that provided the bulk of the diet, women's contribution to the total weight of food collected being between 60 and 80 per cent of the total. The contribution was not just about total amount; plant foods and small game were also more reliable resources than the dispersed and highly mobile large game. Men may have brought back a kangaroo occasionally but the women provided the reliable sustenance.

Women would frequently gather the seeds of grasses and acacias, which they harvested in wooden dishes which were taken back to the campsite, wind-winnowed and separated from the chaff. The seeds were then ground and mixed with water, making a paste which was eaten raw or cooked. Much of the food preparation and cooking activity, using readily available firewood, was done by the women. While the group typically ate one main meal a day, in the late afternoon, everyone ate snacks during the day while hunting or gathering. Favourite snacks were fruit, nectar, tree gum, eggs, fungi, lerp,[15] and honey ants. The youngsters, typically impatient like all children, would eat all they hunted or gathered on the spot, frequently capturing lizards from an early age and skilfully catching birds by throwing sticks and stones at them.

Being nomadic and lacking any beasts of burden, the Mardu were limited by what they could carry with them, especially when changing campsites. Their toolkit could be divided into three basic categories:[16] multi-purpose, appliances, and instant tools. Multi-purpose tools were lightweight and easily carried, mostly made from wood. They included some essential items, like the digging sticks that women used, supplemented by small digging dishes. Dishes were carved, usually by men, from eucalypt wood and bark; the latter are less durable. The larger dishes carried food, water, and even small babies. Women also carried a small stone, which was used against large and flat stones or against flat rock surfaces to grind seeds and other foods.

Men had a larger kit that included wooden throwing spears and the multi-purpose spear-thrower which was used not just to launch spears but also as a tray in which to mix native tobacco and ashes or ochre used in body decoration. The spear-thrower was also used in making fire, as a scraper and knife in woodworking and in the preparation and butchery of game, as a percussion instrument, and even as a hook to get fruits and berries that were out of reach. The famous returning boomerang was not used for hunting but was instead a fighting weapon that doubled up as musical instrument. The only stones that men carried were small flake knives which were used in a variety of cutting and scraping tasks. Most were hand-held but, sometimes, a favourite piece was given a small resin handle. The desert people were clearly adept at making tools that could serve many purposes, thus avoiding having to carry too many things.

The second category was appliances, tools which were normally left at a site and reused on subsequent visits. This means that the Mardu would leave the heavier items behind in camps or particular locations which served specific functions, in the confidence that they would be there when they returned months later. Appliances included the large and heavy base stones that were used in grinding food and ochre, hand-held stone pounders used for mashing bones to get to the marrow, hand axes which weighed between ~2 and 3.5 kilograms (4 and 8 pounds), and a reserve supply of stones for future visits.

Instant tools, the third category, were made from raw materials that were readily available and close at hand. They were used when the need arose and were discarded once finished with. Such tools included stone pounders, axes and flakes, the grass circlets that women made to cushion dishes carried on their heads, some objects used in fire-making, and spindles.

With a body adapted for mobility, an impressive ecological knowledge which offered a varied diet,[17] and a combination of multi-purpose, lightweight tools with heavy items left in select locations, the Mardu were desert experts. Their expertise went beyond the strict ecological knowledge; it also included a deep understanding and an ability to predict the weather. Their vocabulary of cloud and rain types and other meteorological phenomena was extensive. They knew how to predict where and when rain would fall and what that rain might produce. The Mardu also recognized distinct seasons of the year which they gave specific names to.[18]

When it rained, the Mardu strategy was to disperse outwards from the main water sources in small groups, towards places with temporary water along the edges of the rain-affected area. Clay pans and pools were the first to be used to collect water and the Mardu made the most of the game attracted to the ephemeral pools. As these pools evaporated, they would retreat towards the more permanent water bodies. This makes sense as a strategy that first exploited water that would evaporate rapidly, while keeping the main sources of water in reserve. This way they maximized their water reserves. The subsequent retreat depended on a range of factors such as the location of known or expected supplies of food, direction of recent rainfall, and the position of known chains of water sources.

The population was dispersed in small groups or bands but this does not mean that they were unsociable. Often, when bands detected each other's smoke on the horizon they would make contact. Sociability was highly valued by the Mardu and they would use contacts to exchange scarce and valued resources. Maintaining friendly relations with neighbours also allowed groups to extend their foraging activities to other territories in times of severe shortage, but this was exceptional as the Mardu tended to hunt and gather in the same general area year after year. Such extensions of territory were met with reciprocal concessions when it was the other group's time to return the favour.

The success of the desert people lay in a system in which they felt no superiority or independence from the forces of Nature. Instead they saw themselves and Nature as part of a wider cosmic order that included also powerful spiritual beings that could be co-opted to change things to their advantage, as in the hunt. This was the Dreaming. This attitude should not be taken to mean that the Mardu did not have an impact on their environment. They did, through burning the land, digging holes, cutting down trees, uprooting shrubs, clearing campsites, digging out wells, and even placing sticks or stones in the forks of trees to warn of something sacred or dangerous nearby. That the land was not transformed in a major way was in part the product of their religious convictions but it was also ruled out by their technology and the fact that they were few in a vast land.

The Mardu, like the people of Puritjarra in the Pleistocene, represent the most extreme adaptation of *Homo sapiens* to water shortage, which is why I regard it as such a great achievement. In Australia, as in other arid regions, they lived at the very edge of the spectrum of climates available to people. Humans managed to colonize the entire continent of Australia by sticking to the tried-and-tested formula of trees/open-spaces/water, often incorporating rocky habitats, as in South Africa and the rocky regions of Middle Earth. If they could live in the arid centre of Australia, then they could live anywhere else on the continent, from the cool conditions of Tasmania in the south to the tropical world of the north.

A complete contrast to life in the desert is provided by the people who settled along the banks of the Murray and Darling Rivers in south-eastern Australia. While the nomadic desert people reciprocated with their neighbours, it seems that those along the Murray River were strongly territorial and exclusive, quite prepared to defend their boundaries with force. The British anthropologist Alfred Radcliffe-Brown met these people in the early 20th century in what he described as the most densely-populated part of Australia before the days of white settlers.[19] The numbers and densities of the river people were 20–40 times greater than those of the non-river people of the same region. The territorial system along the Murray–Darling Basin may go back a considerable time.

The people living in the area prior to and during the last Ice Age probably hunted the megafauna (e.g. large kangaroos) around lakes and on the plains. They were hunters and gatherers who probably reciprocated and traded with neighbouring groups just like the Mardu did in the arid desert. At the end of the Ice Age, around 21/22 thousand years ago, the landscape changed, with slower-flowing rivers and a change from grasslands to savannah; the megafauna was extinct.[20] People started to bury their dead in formal graveyards and some of the large cemeteries of later times, such as those at Kow Swamp and Coobool Creek, began to be used—an indication of a direct association with the land. This link becomes firmly established after 9 thousand years ago, when there is a notable decrease in the size of the people buried in cemeteries, and a high incidence of cranial trauma, in both men and women, which is indicative of violence. This violence has been linked to the defence of territory.

The population was large, especially when compared to other regions of Australia. A burial site in Lake Victoria contained 10 thousand individuals, giving an idea of the levels of population that the rivers and wetlands of south-eastern Australia held. Cemeteries were established along the banks of the rivers, indicating territoriality based on descent as people defended a land that was a rich larder of fish, fowl, and invertebrates; the river margins contrasted with the barren, parched land of the nearby drought-stricken plain. During drought the river was the only reliable lifeline for food and water and access to these resources was controlled by the resident groups. Ritual practices, notably skull deformation and tooth avulsion,[21] were ways of physically demonstrating identity.

One result of territoriality was genetic isolation. The anatomy of the people of the Murray–Darling Basin was different from that of any other group of Australians, the product of inbreeding. The entire population along the river system was also very diverse which reflected the linear nature of the territories that promoted isolation between the Murray–Darling groups themselves. At one level the river people were isolated from the people of the arid lands beyond and shared features in common. At another level, those from different stretches of river were different from each other, too.

Australia has offered us a cameo of *Homo sapiens* as a desert-dweller, the end point of the 2-million-year-old evolution of the improbable primate. In the people of the deserts of the interior we find the ultimate performance of a species whose ancestry was the rainforest. We have also seen how reciprocity and sharing, a caring and sustainable approach to land, and resource management were a part of the improbable primate's make-up. But we have also glimpsed here the dark side: when living in a resource-rich land, the population rapidly increased and took on an inward-looking and violent streak as they jealously safe-guarded these resources. In this way, the paradoxical nature of *Homo sapiens*, the improbable primate, is painted for us by archaeology and ethnography. We now need to contemplate the canvas to see the improbable primate within us.

11

From Lake Chad to Puritjarra and Beyond

By the 1930s, the Puritjarra Rock Shelter in the Australian desert is silent. For more than 30 thousand years people have lived their lives here. Now they have moved to mission stations and ration depots in the West MacDonnell Ranges.[1] They no longer have to worry about where to go and look for food each day as it has all been organized for them. Their stomachs may be full but they weep for the loss of the free life in the wilderness. The ancestors of the settlers had long lost that freedom and, sadly, the settlers themselves could no longer recognize it when they had it in front of them. With the forced conversion of some of the last surviving populations of hunter-gatherers into civilization, humanity lost its soul.

The 6-million-year-long story that had started on the shores of an ancient wetland in the heart of Middle Earth was over; it was replaced by a much younger story which originated some

10 thousand years ago along the banks of another wetland, also in the very core of Middle Earth.[2] In the dry desert, the fruiting trees have been replaced by springs and fleeting ponds. Most of the time they are not worth defending and the people of the desert move about in groups, using their vision and brains to locate and relocate places with fresh water across a vast waterless landscape.

Part of the primate way of life involves a close bonding with the natal area.[3] The attachment applies just as much to nomadic groups as it does to sedentary ones. The nomads of the Australian desert were attached to their natal home range, which could be vast, and only left it in times of hardship when a system of reciprocity between neighbouring groups allowed for temporary encroachment into adjacent lands. The conditions—highly-dispersed and hard-to-defend resources—promoted reciprocity in favour of aggression and territorial defence. This was not the case in the Murray–Darling Basin. Here resources could be defended and they were to the point of violence as seen by the injuries that many of the people buried in cemeteries endured.

The key point here is that we can end up with a strongly territorial system or a much more egalitarian one simply by changing the conditions. In the case of the Australian Aboriginal comparison it was the distribution and predictability of water which in turn determined the distribution of other valued resources. The people were the same in every respect. What differed were the conditions and the response to these circumstances. The other point is that the observed behaviours were only different from those of the Miocene forest primates in degree. Nomadic or territorial life in *Homo sapiens* arose from a fruit-eating, Miocene rainforest ancestry.

We often associate large-scale territorial behaviour in humans with the advent of production systems, agriculture in particular, but this was clearly not a prerequisite. Agriculture emerged within a specific set of conditions and is a special case of sedentary life but, as the people of the Murray–Darling Basin show us, it was not a prerequisite of large-scale territorial life.

There are other similar examples of sedentary behaviour without agriculture and we ought to recall these briefly here. Twenty-three thousand years ago the people of Ohalo, by the Sea of Galilee, lived in a village of brush huts on the shores of a lake; the rich and predictable resources of the lake permitted a sedentary existence. Here people harvested wild cereals and made baskets from plant fibres. The semi-nomadic Gravettians of eastern and central Europe in the build-up to the last Ice Age around 30 thousand years ago had also found a predictable and abundant food in the form of migrating herds of reindeer; they too established villages, like Pavlov and Dolni Vestonice, and even went down the route of industrial production in the form of clay figurines. The difference between the Gravettians and the nomadic people of Puritjarra is that the former had a predictable and abundant resource while the latter did not. The Natufian people of the Middle East of around 13 thousand years ago were able to become sedentary because they lived in a place of great resource diversity, with seasonal concentrations of gazelles, cereals, and other foods. In this case diversity of resources made for a predictable environment. With the rising sea levels around 8 thousand years ago, people on the north-west Pacific coast of America settled and defended territories against neighbours. They lived in hierarchical societies with a division of labour among their people; the secret in this case was the massive and predictable seasonal arrival of salmon. So much fish was caught in a short time that techniques were developed to preserve and store the surplus. In coastal Peru, the inhabitants set up villages that exploited the abundance of fish along with resources inland and they moved their villages up and down the coast as they followed the fish, in a fashion not dissimilar to the Gravettians.

These examples are all fairly recent in our story, after 30 thousand years ago. Can we find evidence of settlement that is earlier? I think that we can and I start with the Neanderthals in Gorham's Cave in Gibraltar. The Neanderthals lived in a land that combined ecological diversity with periods of superabundance. The latter was provided by the sea in the form of migrations of tuna and the seasonal arrival of monk seals to breed.[4] We have looked at a number of caves and rock shelters that were occupied by the Neanderthals in Gibraltar[5] and we looked at their stone tools for evidence of mobility: where were the raw materials obtained, and were people staying in the caves for a while or just paying short visits? The answers suggested that the Neanderthals were probably behaving like the Gravettians or the Peruvian coastal people, albeit operating on a smaller spatial scale. The furthest raw materials came from a distance of 17 kilometres, not very far for a human to travel. The evidence from the caves along the 6-kilometre stretch of limestone cliffs of Gibraltar was that the Neanderthals were going into them occasionally and briefly but they were not living inside them—not for very long anyway. The difference was Gorham's, which was like the mother cave. Here the Neanderthals lived in a more permanent and sustained manner, akin to a permanent campsite. So, where there was ecological diversity and abundance, the Neanderthals too opted for a sedentary existence. What we do not know is how territorial they were, but we can imagine that they must have been so.

We can only speculate about situations from even earlier times. What were the people of GBY doing 780 thousand years ago (Chapter 6)? Or indeed those early people with their Oldowan toolkits by the lakesides at Ain Hanech, Baza, or Ubeidiya? In all probability they were not moving around more than they needed to. The common denominator, along with predictable resources, in all these sites is the presence of a substantial and fairly permanent water supply. Predictable water argued in favour of a sedentary life but its distribution, I will now argue, restricted the spread of this way of life.

Take the case of the Murray–Darling Basin. People adopted a sedentary life and the population grew to a significant size, but it was contained. It did not spread beyond the boundaries of the river system because the desert lay beyond and, even if people colonized such areas of instability, they would have again been limited to low densities. In other cases, the barrier was imposed by time and changing climatic conditions. The Natufians seem to have been able to domesticate rye but the instability brought by a period of cold and arid conditions around 12.8 thousand years ago[6] ended the experiment. A permanent source of water and stable conditions of resource diversity or abundance were therefore needed for population growth and settlement and it happened on a number of occasions in different parts of the world. Geographical expansions needed something else: a physical continuity of the conditions long enough to permit the spread. That also happened repeatedly, probably on more occasions than we realize and, as we have seen, followed two major avenues: water bodies and high ground, also associated with reliable water supply. Some readers may be surprised that I have not added the coast as an avenue. It is because I do not consider the coast to have provided the conditions suitable for prolonged geographical expansion. At best the coast was a particular example of an avenue that followed water bodies and quite often, as in southern Africa and south-western Iberia, it was a refuge and cul-de-sac, but rarely a launchpad.

These geographical expansions started with the very origin of *Homo sapiens*, 1.8 million years ago. On a number of occasions they conquered significant tracts of geography. The expansions probably did not always involve nomadic people. My guess is that they most often involved sedentary or semi-sedentary people. These would have been the ones that were able to settle in areas of stability and generate population growth spurts which translated into dispersal. The nomadic people would have always been at low density so their rate of expansion would have been slower; when it happened, it would have got people across arid and semi-arid regions.[7] Global expansions probably involved a mix of sedentary and nomadic lifestyles, depending on where the growing populations were situated and what the ecological conditions were like.

The 6-million-year-long story of our evolution is in contrast with the much younger story which started after 10 thousand years ago. This younger story was to lead to the eventual encounter of western colonists and native peoples across Australia and the Americas in particular. Can we see anything in this more recent story that we could relate to the patterns that we have established in this book? I think that we can. If we take a broad survey of regions in which agriculture originated or had an early success,[8] we find that they have features in common. For starters they all occurred in the time frame after 10 thousand years ago, which is when the world reached a level of stability in the context of a warm interglacial, the Holocene. Many of these areas of origin or early spread are associated with significant water bodies, usually large rivers in warm, often arid, areas of Middle Earth and beyond: Tigris and Euphrates Rivers; Jordan River; Yangtze and Yellow Rivers; Indus River; Nile River; Mississippi and Ohio Rivers; and possibly the Niger River in the African Sahel. A second group is associated with hills and mountains in tropical areas, which as we saw provided alternative conditions as refuges and launchpads in earlier times: the Andes; Meso-American Highlands; New Guinea Highlands; and the Ethiopian Highlands. One region in which agriculture did not originate but to which it spread rapidly was western Europe, particularly around the warm Mediterranean. We have already seen how parts of the oceanic west of the Eurasian continent had been favoured by the Neanderthals on account of the generous rainfall there. That rainfall provided the predictable and stable conditions that equated to the large rivers or the tropical highlands.

We have seen in this book how the favoured human habitat combined trees, open spaces, and water. The farmers developing agriculture in some areas, such as the New Guinean Highlands or western Europe, hit areas of dense forest. They resolved the problem by opening these up by cutting down or burning trees, later by allowing grazing animals like goats to maintain a patchwork of open spaces and trees in places with abundant water. The active land management that started in Niah Cave and in Australia was now taken to a new level. Where the ideal human habitat was not available, we created it—provided there was water at hand. This powerful combination of increasing the environment's carrying capacity by growing and farming our favourite foods, linked to intense land management, created an unprecedented population explosion. Now we were no longer hemmed in. Provided there was water, and we found systems of harnessing those waters for irrigation, we could spread across the world in an unprecedented manner, and we did.

Much later, in Renaissance Italy, or in the Age of Enlightenment, or in Victorian England, painters chose to portray the ideal and idyllic human landscape in their works of art. Almost without exception these wonderful works of art featured certain elements that will by now be familiar to us: trees, open spaces, and water. Many also included grazing herbivorous animals—cows, horses, sheep—which were domesticated equivalents of the kinds of wild animals that we grew up with during the course of our evolution. They also included rocky places or their modern equivalents, buildings.[9] In the high-density, urban, post-Industrial Revolution world of the west, we created parks inside cities. They too always combine trees, open spaces, and water. The improbable primate, long-distance runner and walker par excellence, now chooses to jog across these parks that recreate the ancestral habitat. We may have taken huge cultural and technological strides in the last 10 thousand years but our biology, put together slowly over millions of years, still shines through no matter how hard we try to pretend otherwise.

ENDNOTES

PREFACE

1. E. Mayr, 'Taxonomic Categories of Fossil Hominids', *Cold Spring Harbor Symposia on Quantitative Biology* 15 (1950): 109–18.
2. J. R. Stewart and C. B. Stringer, 'Human Evolution Out of Africa: The Role of Refugia and Climate Change', *Science* 335 (2012): 1317–21.
3. E. Mayr, *Animal Species and Evolution* (Cambridge, Mass.: Harvard University Press, 1963).
4. E. Aguirre, '*Homo erectus* and *Homo sapiens*: One or More Species?', *Courier Forschungs-Institut Senckenberg* 171 (1994): 333–9; M. H. Wolpoff, 'The Systematics of *Homo*', *Science* 284 (1999): 1773c.
5. G. G. Simpson, 'The Species Concept', *Evolution* 5 (1951): 285–98. Palaeontological species are all fossil species that intergrade to form a chronological series. They are defined by arbitrary morphological limits in the time series. They are, in effect, lineages. Whether two lineages represent distinct biological species, i.e. they could not have interbred, or not is impossible to determine on morphological grounds alone. Recent advances in genetics, including fossil DNA, are resolving some of these issues. See also, A. J. Cain, *Animal Species and Their Evolution* (London: Hutchinson, 1954).

6. C. Finlayson, *The Humans Who Went Extinct: Why Neanderthals Died Out and We Survived* (Oxford: Oxford University Press, 2009).
7. C. Finlayson, *Neanderthals and Modern Humans: An Ecological and Evolutionary Prespective* (Cambridge: Cambridge University Press, 2004).
8. F. L. Mendez et al., 'An African American Paternal Lineage Adds an Extremely Ancient Root to the Human Y Chromosome Phylogenetic Tree', *American Journal of Human Genetics* 92 (2013): 454–9.
9. D. Lordkipanidze et al., 'A Complete Skull from Dmanisi, Georgia, and the Evolutionary Biology of Early *Homo*', *Science* 342 (2013): 326-31.

CHAPTER 1

1. Finlayson, *The Humans Who Went Extinct*.
2. K. Milton, 'Diet and Primate Evolution', *Scientific American* (Aug. 1993): 86–94.
3. Milton studied the highly active fruit-eating black-handed spider monkeys *Ateles geoffroyi* on Barro Colorado Island in Panama. There, she compared them with the less-active mantled howler monkeys *Alouatta palliata* which took a greater proportion of leaves and flowers and fewer fruit than the spider monkeys. Her study was the basis for future work on the relationship between diet and behaviour in primates. K. Milton, 'Food Choice and Digestive Strategies of Two Sympatric Primate Species', *American Naturalist* 117 (1981): 496–505.
4. Among the most intensely-studied primates and our closest living relatives. For range of foods consumed see R. Wrangham, 'Feeding Behaviour of Chimpanzees in Gombe National Park,

Tanzania', in T. H. Clutton-Brock (ed.), *Primate Ecology* (London: Academic Press, 1977).

5. See K. B. Strier, *Primate Behavioral Ecology* (Boston: Allyn and Bacon, 2000) for a useful summary.
6. J. L. Gittleman (ed.), *Carnivore Behavior, Ecology, and Evolution* (Ithaca, NY: Cornell University Press, 1989), vol. i.
7. Based on cranial shape analysis. B. Figueirido, F. J. Serrano-Alarcón, G. J. Slater, and P. Palmqvist, 'Shape at the Cross-roads: Homoplasy and History in the Evolution of the Carnivoran Skull towards Herbivory', *J. Evolutionary Biol.* 23 (2010): 2579–94; B. Figueirido, P. Palmqvist, J. A. Pérez-Claros, and W. Dong, 'Cranial Shape Transformation in the Evolution of the Giant Panda (*Ailuropoda melanoleuca*)', *Naturwissenschaften* 98 (2011): 107–16.

CHAPTER 2

1. A feature—the ability to eat a wide range of foods—that allowed primates to occupy many different ecological opportunities. I. Crowe, *The Quest for Food: Its Role in Human Evolution and Migration* (Stroud: Tempus, 2000).
2. Whiten et al. summarized information from seven long-term chimpanzee *Pan troglodytes* study sites and found that 39 different behaviour patterns, including tool use, courtship, and grooming, were present in some communities and not in others. They discounted ecological explanations for the differences which they attributed to cultural variations. A. Whiten et al., 'Cultures in Chimpanzees', *Nature* 399 (1999): 682–5.
3. The time when the earliest fossils of our potential ancestors are thought to have lived. These have been ascribed to *Sahelanthropus tchadensis.* M. Brunet et al., 'A New Hominid from the Upper

Miocene of Chad, Central Africa', *Nature* 418 (2002): 145–51. I discuss these fossils' relevance to our ancestry in *The Humans Who Went Extinct.*

4. The evidence for these dramatic changes in the environment is described in *The Humans Who Went Extinct* so I will not elaborate further. Since *Humans,* Cerling has provided a useful analysis of environmental conditions in eastern Africa over the past 6 million years (hereafter myr), supporting the view that grasslands became dominant during this time. This was indeed a new world of grasslands and the world changed for ever. T. E. Cerling et al., 'Woody Cover and Hominin Environments in the Past 6 Million Years', *Nature* 476 (2011): 51–6.
5. The recent discussion has revolved around the habitat of *Ardipithecus ramidus* which is the best known of these early ancestors and lived in Ethiopia around 4.51–4.32 myr: T. D. White et al., 'Macrovertebrate Paleontology and the Pliocene Habitat of *Ardipithecus ramidus*', *Science* 326 (2009): 67, 87–93; T. E. Cerling et al., 'Comment on the Paleoenvironment of *Ardipithecus ramidus*', *Science* 328 (2010): 1105d; T. D. White et al., 'Response to Comment on the Paleoenvironment of *Ardipithecus ramidus*', *Science* 328 (2010): 1105e.
6. See reasoning behind lack of credence of the Aquatic Ape Theory in John Hawks's weblog, 'Why anthropologists don't accept the Aquatic Ape Theory', at <http://www.johnhawks.net/weblog/topics/pseudoscience/aquatic_ape_theory.html>. There is also a detailed critique by Jim Moore, 'Aquatic Ape Theory: Sink or Swim?', <http://www.aquaticape.org/>.
7. A. Hardy, 'Was Man More Aquatic in the Past?', *New Scientist*, 17 March 1960.

8. See, for example, E. Morgan, *The Aquatic Ape: A Theory of Human Evolution* (New York: Stein and Day, 1982); E. Morgan, *The Aquatic Ape Hypothesis* (Bury St Edmunds: Souvenir Press, 1997).
9. Toumaï and Ramidus were vernacular names which I used in *The Humans Who Went Extinct*.
10. P. Vignaud et al., 'Geology and Palaeontology of the Upper Miocene Toros-Menalla Hominid Locality, Chad', *Nature* 418 (2002): 152–5.
11. *Anancus kenyensis* and *Sivatherium spp.*, Vignaud et al., 'Geology and Palaeontology'.
12. Isotopes are chemical variants of an element, sharing the same number of protons but differing in the number of neutrons. The ratio of different isotopes in teeth can give a good indication of diet. Carbon isotopes are taken into the teeth through the diet during the course of an animal's life. We can infer, for example, if the diet of a herbivorous animal consisted largely of grazing grasses (C_4 plants) or browsing leaves (C_3 plants). In the case of Ramidus, the carbon isotope signal of its teeth resembled that of the small arboreal baboon *Pliopapio* and the leaf-browser *Tragelaphus* (see also nn. 14 and 15) suggesting little intake of grasses. See also *The Humans Who Went Extinct*, 76–7.
13. The relevant papers in the *Science* volume are referred to in n. 5. See also: T. White et al., '*Ardipithecus ramidus* and the Paleobiology of Early Hominids', *Science* 326 (2009): 64; G. Woldegabriel et al., 'The Geological, Isotopic, Invertebrate, and Lower Vertebrate Surroundings of *Ardipithecus ramidus*', *Science* 326 (2009): 65; A. Louchart et al., 'Taphonomic, Avian, and Small-Vertebrate Indicators of *Ardipithecus ramidus* Habitat', *Science* 326 (2009): 66.

14. *Kuseracolobus aramsi* which was a leaf-eater with a strong preference for trees. The carbon isotope signature of its teeth indicated that it was feeding in dense to open forest. The third primate was a small baboon, *Pliopapio alemui.*
15. A kudu, *Tragelaphus* species, whose present-day counterparts are associated with wooded habitats.
16. Dukiers, *Cephalophus* species, are small antelopes that are common in the dense African forests.
17. Arguing for a tree or bush savannah with 25 per cent or less woody canopy cover. The environment ranged from riparian forest to grassland with C_4 grasses making up between 40 and 60 per cent of the biomass. Cerling et al., 'Comment on the Paleoenvironment of *Ardipithecus ramidus*'.
18. Of 275 fish specimens, catfish *Clarias* dominated (175 specimens) followed by barbs *Barbus* (20) and cichlids *Cichlidae* (21). These are species of poorly oxygenated, shallow waters. The crocodiles were indistinguishable from present-day Nile crocodiles. Freshwater turtles were the African mud turtle *Pelusios* and a flapshell softshell turtle *Pelomedusa.* Woldegabriel et al., 'The Geological, Isotopic, Invertebrate, and Lower Vertebrate Surroundings'.
19. The scavenging habits of wild boar in southern Spain have been the subject of a recent study: E. Bernáldez Sánchez, 'Biostratinomy Applied to the Interpretation of Scavenger Activity in Paleoecosystems', *Quaternary International* 243 (2011): 161–70.
20. Whittaker gives a list of the productivity of world biomes (in grams of organic matter per square metre per year). Tropical rainforest is second after algal beds and reefs with 2,200 g/m^2 and it is followed by swamp and marsh (2,000) and tropical seasonal forest (1,600). Savannah is eighth at 900 g/m^2.

R. H. Whittaker, *Communities and Ecosystems* (New York: Macmillan, 1975).

21. This paragraph, taken from Cerling et al.'s paper ('Comment on the Paleoenvironment of *Ardipithecus ramidus*'), shows how predefined habitat categories can easily get us tied down in a level of detail that the initial observations simply cannot answer and in debates that cannot be resolved: 'We adopt the United Nations Scientific and Cultural Organization (UNESCO) definitions for classification of African vegetation used by White et al. ['Macrovertebrate Paleontology'] in which "forests have continuous stands of trees with overlapping crowns, forming a closed, often multistory canopy 10 to 50 m high; the sparse ground layer usually lacks grasses; closed woodlands have less continuous canopies and poorly developed grass layers; woodlands have trees with canopy heights of 8 to 20 m; their crowns cover at least 40 per cent of the land surface but do not overlap extensively. Woodland ground layer always includes heliophilous (sun-loving, C_4) grasses, herbs/forbs, and incomplete small tree and shrub understories; scrub woodland has a canopy height less than 8 m, intermediate between woodland and bushland. As proportions of bushes, shrub, and grasses increase, woodlands grade into bushland/thickets or wooded grasslands. Wooded grassland is land covered with grasses and other herbs, with woody plants covering between 10 and 40 per cent of the ground. Grassland is land covered with grasses or other herbs, either without woody plants or the latter not covering more than 10 per cent of the ground." The UNESCO classification does not include a definition of savanna because of its ambiguous use, but common usage of the term would

include wooded grasslands and grasslands of the UNESCO terminology.'

22. Facultative bipedal means that it could walk on the two hind limbs but that its anatomy did not make it exclusively (obligate) bipedal. Since it was partly arboreal we should conclude that Ramidus walked on two hind legs for part of the time and on all fours at other times, perhaps when on the branches of trees. We have little to inform us on Toumaï's level of bipedality.
23. I discuss this method of walking on trees by orang-utans in *Humans Who Went Extinct*, 35.

CHAPTER 3

1. Barbary macaques *Macaca sylvanus*, the famous rock apes, roam freely on Gibraltar. They were introduced from North Africa in the 18th century by the British. They live in troops and defend territories. The Gibraltar Government provides food and water for them in fixed locations but groups splinter and separate away from these sources, often raiding houses in the town below. They also forage on wild foods. During the summer in particular, these rogue groups do not have access to fixed water points.
2. Australopithecines, the species of early hominids classified under the genera *Australopithecus* and *Paranthropus*, that lived in Africa between 4.2 and 1.4 myr. I refer to them as proto-humans in *The Humans Who Went Extinct*.
3. *The Humans Who Went Extinct*, table 1, p. 26.
4. To the species listed in table 1 in *The Humans Who Went Extinct* we must add the new species—*Australopithecus sediba*—from Malapa in South Africa; R. Pickering et al., '*Australopithecus sediba* at 1.977 Ma and Implications for the Origins of the

Genus *Homo*', *Science* 333 (2011): 1421–3. In this book I will refer to this species as *sediba*.

5. The summaries are K. Reed, 'Early Hominid Evolution and Ecological Change through the African Plio-Pleistocene', *Journal of Human Evolution* 32 (1997): 289–322, and B. Wood and D. Strait, 'Patterns of Resource Use in Early *Homo* and *Paranthropus*', *Journal of Human Evolution* 46 (2004): 119–62.
6. Grasslands that are partially flooded or subject to regular flooding.
7. A good recent summary is P. Ungar and M. Sponheimer, 'The Diets of Early Hominins', *Science* 334 (2011): 190–3. Dental microwear patterns—usually pits and scratches in varying combinations—are detected in the teeth of fossil hominids. They reflect what an individual was eating weeks before its death, the 'Last Supper' phenomenon. Hard and brittle foods, such as nuts, produce dental microwear patterns that are dominated by pits and complex patterns on the tooth surface. Soft but tough foods, such as leaves, produce scratched surfaces which are typically seen as parallel lines. Surface patterns are relatively simple. Combinations of the two patterns among individuals of the same species would reflect catholic food choice relative to specialized species. Stable isotopes, especially carbon isotopes, are represented in the growth layers of teeth and reflect diet through time. In studies of early hominids they have been especially useful in identifying the proportion of C_3 and C_4 plants consumed. See also Ch. 2 n. 12.
8. Summarized in Ungar and Sonheimer, 'The Diets of Early Hominins', with additonal summary and data on two individuals of *A. sediba* in A. G. Henry et al., 'The Diet of *Australopithecus sediba*', *Nature* 487 (2012): 90–3.

9. F. E. Grine et al., 'Molar Microwear in *Praeanthropus afarensis*: Evidence for Dietary Stasis through Time and under Diverse Paleoecological Conditions', *Journal of Human Evolution* 51 (2006): 297–319.
10. Ungar and Sonheimer, 'The Diets of Early Hominins'; Henry et al., 'The Diet of *Australopithecus sediba*'; Grine et al., 'Molar Microwear in *Praeanthropus afarensis*'.
11. Henry et al., 'The Diet of *Australopithecus sediba*'.
12. From the Greek, meaning 'plant stone'. Plants take up silica from the soil which is then deposited in various parts of the plant—the phytoliths. They are tough microscopic structures which can be preserved. The study of plant phytoliths is not new but finding them in the teeth of Neanderthals was a first for science (A. G. Henry et al., 'Microfossils in Calculus Demonstrate Consumption of Plants and Cooked Foods in Neanderthal diets (Shanidar III, Iraq; Spy I and II, Belgium)', *Proceedings of the National Academy of Sciences USA* 108 (2011): 486–91). Amanda Henry's work then looked at earlier hominids and she recovered 38 phytoliths from Sediba.
13. M. Sponheimer et al., 'Hominins, Sedges, and Termites: New Carbon Isotope Data from the Sterkfontein Valley and Kruger National Park', *Journal of Human Evolution* 48 (2005): 301–12.
14. White et al., '*Ardipithecus ramidus* and the Paleobiology of Early Hominids'; B. Latimer and C. O. Lovejoy, 'Hallucal Tarsometatarsal Joint in *Australopithecus afarensis*', *American Journal of Physical Anthropology* 82 (1990): 125–33.
15. K. Wong, 'Footprints to Fill: Flat Feet and Doubts about Makers of the Laetoli Tracks', *Scientific American* 293 (2005): 18–19.
16. T. L. Kivell et al., '*Australopithecus sediba* Hand Demonstrates Mosaic Evolution of Locomotor and Manipulative Abilities',

Science 333 (2011): 1411–17; B. Zipfel et al., 'The Foot and Ankle of *Australopithecus sediba*', *Science* 333 (2011): 1417–20.

17. *The Humans Who Went Extinct*, table 1, p. 26.
18. Seventeen of the 21 sites (80.9 per cent) had trees and 19 of the 21 sites (90.5 per cent) had open spaces. In contrast, only 10 (47.6 per cent) had bushes.
19. Reed, 'Early Hominid Evolution and Ecological Change'.
20. Ungar and Sponheimer, 'The Diets of Early Hominins'. See also Ch. 2 n. 12.
21. B. Wood and M. Collard, 'The Human Genus', *Science* 284 (1999): 65–71.
22. Eight of nine sites record water, eight of nine trees, and all nine open spaces.
23. P. deMenocal, 'African Climate Change and Faunal Evolution during the Pliocene-Pleistocene', *Earth and Planetary Science Letters* 220 (2004): 3–24.
24. T. E. Cerling et al., 'Woody Cover and Hominin Environments in the Past 6 million Years', *Nature* 476 (2011): 51–6.
25. T. E. Cerling, 'Development of Grasslands and Savannas in East Africa during the Neogene', *Palaeogeography, Palaeoclimatology, Palaeoecology* 97 (1992): 241–7.
26. S. C. Reynolds, G. N. Bailey, and G. C. P. King, 'Landscapes and Their Relation to Hominin Habitats: Case Studies from *Australopithecus* Sites in Eastern and Southern Africa', *Journal of Human Evolution* 60 (2011): 281–98.
27. S. Semaw et al., '2.6 Million-Year-Old Stone Tools and Associated Bones from OGS-6 and OGS-7, Gona, Afar, Ethiopia', *Journal of Human Evolution* 45 (2003): 169–77. See also *Humans Who Went Extinct*, 41 and 226 n. 28.

28. M. Haslam et al., 'Primate Archaeology', *Nature* 460 (2009): 339–44.
29. S. P. McPherron et al., 'Evidence for Stone-Tool-Assisted Consumption of Animal Tissues before 3.39 Million Years Ago at Dikika, Ethiopia', *Nature* 466 (2010): 857–60.
30. P. S. Ungar (ed.), *Evolution of the Human Diet: The Known, the Unknown and the Unknowable* (Oxford: Oxford University Press, 2007).

CHAPTER 4

1. F. Brown et al., 'Early *Homo erectus* Skeleton from West Lake Turkana, Kenya', *Nature* 316 (1985): 788–92.
2. Finlayson, *The Humans Who Went Extinct*, 54.
3. *Homo erectus* always incorporated trees, open spaces, and water (all seven sites examined). Bushland was only present in three (42.9 per cent) sites and rocky habitats in two (28.6 per cent). The proportions are remarkably similar to those for the australopithecines.
4. deMenocal, 'African Climate Change'.
5. R. R. Graves et al., 'Just How Strapping was KNM-WT 15,000?', *Journal of Human Evolution* 59 (2010): 542–54.
6. H. M. McHenry, 'Femoral Length and Stature in Plio-Pleistocene Hominids', *American Journal of Physical Anthropology* 85 (1991): 149–58.
7. Estimated body weight of *Homo erectus*: 63 kg (139 lb, male) and 52 kg (114 lb, female). Compare with *habilis*: 52 kg (114 lb, male) and 32 kg (70 lb, female). From H. M. McHenry, 'Behavioral Ecological Implications of Early Hominid Body Size', *Journal of Human Evolution* 27 (1994): 77–87.
8. Energy expended per unit weight per distance moved.

9. R. H. Peters, *The Ecological Implications of Body Size* (Cambridge: Cambridge University Press, 1983).
10. K. L. Steudel-Numbers, 'Energetics in *Homo erectus* and Other Early Hominins: The Consequences of Increased Lower-Limb Length', *Journal of Human Evolution* 51 (2006): 445–53.
11. K. L. Steudel-Numbers and M. J. Tilkens, 'The Effect of Lower Limb Length on the Energetic Cost of Locomotion: Implications for Fossil Hominins', *Journal of Human Evolution* 47 (2004): 95–109.
12. K. L. Steudel-Numbers et al., 'The Evolution of Human Running: Effects of Changes in Lower-Limb Length on Locomotor Economy', *Journal of Human Evolution* 53 (2007): 191–6.
13. Antelopes, horses, elephants, rhinos. See, for example, T. G. Bromage and F. Schrenk (eds.), *African Biogeography, Climate Change, & Human Evolution* (New York: Oxford University Press, 1999).
14. Cats, dogs, and hyenas. See A. Turner, 'The Evolution of the Guild of Larger Terrestrial Carnivores during the Plio-Pleistocene in Africa', *Geobios* 23 (1990): 349–68; L. Werdelin and M. E. Lewis, 'Plio-Pleistocene Carnivora of Eastern Africa: Species Richness and Turnover Patterns', *Zoological Journal of the Linnean Society* 144 (2005): 121–44.
15. B. R. Benefit, 'Biogeography, Dietary Specialization, and the Diversification of African Plio-Pleistocene Monkeys', in Bromage and Schrenk (eds.), *African Biogeography*, 172–88.
16. T. D. White, 'African Omnivores: Global Climatic Change and Plio-Pleistocene Hominids and Suids', in E. S. Vrba et al. (eds.), *Paleoclimate and Evolution with Emphasis on Human Origins* (New Haven: Yale University Press, 1995), 369–84.

17. In southern Europe at least, wild boar *Sus scrofa* scavenge carcasses in a similar fashion to hyenas. E. Bernáldez Sánchez, 'Biostratinomy Applied to the Interpretation of Scavenger Activity in Paleoecosystems', *Quaternary International* 243 (2011): 161–70.
18. J. M. Baldwin, 'A New Factor in Evolution', *American Naturalist* 30 (1896): 441–51 and 536–53; but the ideas were independently conceived by Lloyd-Morgan and Osborn: C. Lloyd Morgan, 'On Modification and Variation', *Science* 4 (1896): 733–40; H. F. Osborn, 'Ontogenic and Phylogenic Variation', *Science* 4 (1896): 786–9.
19. P. Bateson, 'The Active Role of Behaviour in Evolution', *Biology and Philosophy* 19 (2004): 283–98.
20. For examples of predator activity patterns see M. Odden and P. Wegge, 'Spacing and Activity Patterns of Leopards *Panthera pardus* in the Royal Bardia National Park, Nepal', *Wildlife Biology* 11 (2005): 145–52; M. W. Hayward and G. Hayward, 'Activity Patterns of Reintroduced Lion *Panthera leo* and Spotted Hyaena *Crocuta crocuta* in the Addo Elephant National Park, South Africa', *African Journal of Ecology* 45 (2007): 135–41.
21. K. Modal et al., 'Response of Leopards to Re-introduced Tigers in Sariska Tiger Reserve, Western India', *International Journal of Biodiversity and Conservation* 45 (2012): 228–36.
22. Specifically cheetahs and wild dogs, B. C. R. Bertram, 'Serengeti Predators and Their Social Systems', in A. R. E. Sinclair and M. Norton-Griffiths (eds.), *Serengeti: Dynamics of an Ecosystem* (Chicago: University of Chicago Press, 1979).
23. S. M. Durant, 'Competition Refuges and Coexistence: An Example from Serengeti Carnivores', *Journal of Animal Ecology* 67 (1998): 370–86.

24. R. W. Newman, 'Why Man Is Such a Sweaty and Thirsty Naked Animal: A Speculative Review', *Human Biology* 42 (1970): 12–27.
25. For some time, the first appearance of the Acheulian and its characteristic hand-axes had been dated to 300 thousand years (hereafter kyr) after the appearance of *Homo erectus*. Two new papers have revised these dates and have shown that the emergence of *Homo erectus* and the appearance of the Acheulian were concurrent around 1.7 myr. C. J. Lepre et al., 'An Earlier Origin for the Acheulian', *Nature* 477 (2011): 82–5; Y. Beyene, 'The Characteristics and Chronology of the Earliest Acheulean at Konso, Ethiopia', *Proceedings of the National Academy of Sciences USA* 110 (2013): 1584–91.

CHAPTER 5

1. C. J. Lepre and D. V. Kent, 'New Magnetostratigraphy for the Olduvai Subchron in the Koobi Fora Formation, Northwest Kenya, with Implications for Early Homo', *Earth and Planetary Science Letters* 290 (2010): 362–74; I. McDougall et al., 'New Single Crystal ^{40}Ar/^{39}Ar Ages Improve Time Scale for Deposition of the Omo Group, Omo-Turkana Basin, East Africa', *Journal of the Geological Society, London* 169 (2011): 213–26.
2. The oldest known Asian *Homo erectus* are two from Mojokerto in Java and are thought to have lived around 1.81 ± 0.04 and 1.66 ± 0.04 myr. C. C. Swisher III et al., 'Age of the Earliest Known Hominids in Java, Indonesia', *Science* 263 (1994): 1118–21.
3. R. Dennell and W. Roebroeks, 'An Asian Perspective on Early Human Dispersal from Africa', *Nature* 438 (2005): 1099–1104.
4. See *Humans Who Went Extinct*, 53–4.

5. L. Gabunia et al., 'Earliest Pleistocene Hominid Cranial Remains from Dmanisi, Republic of Georgia: Taxonomy, Geological Setting, and Age', *Science* 288 (2000): 1019–25.
6. Finlayson, *Neanderthals and Modern Humans.*
7. Dennell and Roebroeks, 'An Asian Perspective on Early Human Dispersal from Africa'.
8. D. Lordkipanidze et al., 'Postcranial Evidence from Early *Homo* from Dmanisi, Georgia', *Nature* 449 (2007): 305–10.
9. J. Hawks, 'News flash: Dmanisi hominids were not short', <http://www.johnhawks.net/weblog/fossils/lower/dmanisi/dmanisi_postcrania_nature_2007.html>.
10. J. Kappelman, 'The Evolution of Body Mass and Relative Brain Size in Fossil Hominids', *Journal of Human Evolution* 30 (1996): 243–76; F. Spoor et al., 'Implications of New Early Homo Fossils from Ileret, East of Lake Turkana, Kenya', *Nature* 448 (2007): 688–91.
11. Finlayson, *The Humans Who Went Extinct.*
12. L. C. Aiello and P. Wheeler, 'The Expensive-Tissue Hypothesis', *Current Anthropology* 36 (1995): 199–221.
13. C. V. Gisolfi and F. Mora, *The Hot Brain: Survival, Temperature and the Human Body* (Cambridge, Mass.: Bradford, 2000).
14. R. B. Eckhardt, 'Was Plio-Pleistocene Hominid Brain Expansion a Pleiotropic Effect of Adaptation to Heat Stress?', *Anthropologischer Anzeiger* 45 (1987): 193–201.
15. Known as the Oldowan.
16. Beyene, 'Characteristics and Chronology'.
17. R. J. Blumenschine and C. R. Peters, 'Archaeological Predictions for Hominid Land Use in the Paleo-Olduvai Basin, Tanzania, during Lowermost Bed II Times', *Journal of Human Evolution* 34 (1998): 565–607; R. Potts et al., 'Paleolandscape

Variation and Early Pleistocene Hominid Activities: Members 1 and 7, Olorgesailie Formation, Kenya', *Journal of Human Evolution* 37 (1999): 747–88; C. Shipton, 'Taphonomy and Behaviour at the Acheulean Site of Kariandusi, Kenya', *African Archaeological Review* 28 (2011): 141–55.

18. S. C. Antón, 'Natural History of *Homo erectus*', *Yearbook of Physical Anthropology* 46 (2003): 126–70.
19. J. C. A. Joordens et al., 'Relevance of Aquatic Environments for Hominins: A Case Study from Trinil (Java, Indonesia)', *Journal of Human Evolution* 57 (2009): 656–71.
20. *Stegodon trigonocephalus.*
21. E. A. Bettis III et al., 'Way Out of Africa: Early Pleistocene Paleoenvironments Inhabited by *Homo erectus* in Sangiran, Java', *Journal of Human Evolution* 56 (2009): 11–24.
22. Antón, 'Natural History of *Homo erectus*'.
23. M. J. Morwood et al., 'Fission-Track Ages of Stone Tools and Fossils on the East Indonesian Island of Flores', *Nature* 392 (1998): 173–6; P. B. O'Sullivan et al., 'Archaeological Implications of the Geology and Chronology of the Soa basin, Flores, Indonesia', *Geology* 29 (2001): 607–10; A. Brumm et al., 'Early Stone Technology on Flores and Its Implications for *Homo floresiensis*', *Nature* 441 (2006): 624–8.
24. O. F. Huffman and Y. Zaim, ' Mojokerto Delta, East Jawa: Paleoenvironment of *Homo modjokertensis*—First Results', *Journal of Mineral Technology* 10 (2003): 1–32.
25. S. Pappu, 'Early Pleistocene Presence of Acheulian Hominins in South India', *Science* 331 (2011): 1596–9.
26. M. Belmaker and E. Tchernov, 'New Evidence for Hominid Presence in the Lower Pleistocene of the Southern Levant', *Journal of Human Evolution* 43 (2002): 43–56.

27. The dates for the site fall within the 1.95–1.77-myr bracket. M. Sahnouni et al., 'Further Research at the Oldowan site of Ain Hanech, North-eastern Algeria', *Journal of Human Evolution* 43 (2002): 925–37.
28. M. Sahnouni et al., 'Ecological Background to Plio-Pleistocene Hominin Occupation in North Africa: The Vertebrate Faunas from Ain Boucherit, Ain Hanech and El-Kherba, and Paleosol Stable-Carbon-Isotope Studies from El-Kherba, Algeria', *Quaternary Science Reviews* 30 (2011): 1303–17.
29. Fossil hominid and Oldowan technology. This is a cave site that is not informative about the ecological conditions outside the cave. Some mammals appear to have been butchered in the cave but the faunal list does not permit us to reconstruct the environment of these hominids. E. Carbonell et al., 'The First Hominin of Europe', *Nature* 452 (2008): 465–70.
30. O. Oms et al., 'Early Human Occupation of Western Europe: Paleomagnetic Dates for Two Paleolithic Sites in Spain', *PNAS* 97 (2000): 10666–70; D. Barsky et al., 'Raw Material Discernment and Technological Aspects of the Barranco León and Fuente Nueva 3 Stone Assemblages (Orce southern Spain)', *Quaternary International* 223–4 (2010): 201–19; I. Toro-Moyano et al., 'The Oldest Human Fossil in Europe Dated to ca 1.4 Ma at Orce (Spain)', *Journal of Human Evolution* (2013): <http://www.dx.doi.org/10.1016/j.jhevol.2013.01.012>.
31. G. R. Scott and L. Gibert, 'The Oldest Hand-Axes in Europe', *Nature* 461 (2009): 82–5.
32. J. Agustí et al., 'The Early Pleistocene Small Vertebrate Succession from the Orce Region (Guadix-Baza Basin, SE Spain) and Its Bearing on the First Human Occupation of Europe', *Quaternary International* 223–4 (2010): 162–9.

33. Gabunia et al., 'Earliest Pleistocene Hominid Cranial Remains from Dmanisi'.
34. R. X. Zhu et al., 'New Evidence on the Earliest Human Presence at High Northern Latitudes in Northeast Asia', *Nature* 431 (2004): 559–62.
35. R. X. Zhu et al., 'Early Evidence of the Genus *Homo* in East Asia', *Journal of Human Evolution* 55 (2008): 1075–85.
36. Dennell and Roebroeks, 'An Asian Perspective on Early Human Dispersal from Africa'.

CHAPTER 6

1. R. Dennell, *The Palaeolithic Settlement of Asia* (Cambridge: Cambridge University Press, 2009).
2. deMenocal, 'African Climate Change'.
3. L. Barham and P. Mitchell (eds.), *The First Africans: African Archaeology from the Earliest Toolmakers to Most Recent Foragers* (Cambridge: Cambridge University Press, 2008).
4. Finlayson, *Neanderthals and Modern Humans.* Although flake technology akin to the Oldowan, which allowed a quick and flexible response to local variability of resources, meant that flakes were still being made into historical times in Africa, Barham and Mitchell (eds.), *The First Africans.*
5. Barham and Mitchell (eds.), *The First Africans.*
6. R. Potts et al., 'Paleolandscape Variation and Early Pleistocene Hominid Activities: Members 1 and 7, Olorgesailie Formation, Kenya', *Journal of Human Evolution* 37 (1999): 747–88.
7. N. Goren-Inbar et al., 'Pleistocene Milestones on the Out-of-Africa Corridor at Gesher Benot Ya'aqov, Israel', *Science* 289 (2000): 944–7; N. Goren-Inbar et al., *The Acheulian Site of*

Gesher Benot Ya'akov, Israel, i. *The Wood Assemblage* (Oxford: Oxbow Books, 2002).

8. *Palaeoloxodon (Elephas) antiquus*. N. Goren-Inbar et al., 'A Butchered Elephant Skull and Associated Artifacts from the Acheulian Site of Gesher Benot Ya'akov, Israel', *Paléorient* 20 (1994): 99–112.
9. Edible plants included wild grape *Vitis sylvestris*, water chestnut *Trapa natans*, prickly water lily *Euryale ferox*, cattail *Typha* spp., oak *Quercus* spp., wild pistachio *Pistacia atlantica*, wild olive *Olea europaea*, plum *Prunus* spp., and jujube *Ziziphus spina-christi*.
10. N. Goren-Inbar et al., 'Nuts, Nut Cracking, and Pitted Stones at Gesher Benot Ya'aqov, Israel', *PNAS* 99 (2002): 2455–60.
11. N. Goren-Inbar et al., 'Evidence of Hominin Control of Fire at Gesher Benot Ya'aqov, Israel', *Science* 304 (2004): 725–7.
12. R. Wrangham, *Catching Fire: How Cooking Made Us Human* (London: Profile Books, 2009).
13. Dennell, *The Palaeolithic Settlement of Asia*.
14. Two classic reviews by Clive Gamble summarize the European Middle Pleistocene: C. Gamble, *The Palaeolithic Settlement of Europe* (Cambridge: Cambridge University Press, 1986); C. Gamble, *The Palaeolithic Societies of Europe* (Cambridge: Cambridge University Press, 1999).
15. Typical European Middle Pleistocene large mammals associated with *Homo erectus* include deer, rhinos, elephants, bison, musk ox, reindeer, wolves, bears, lions, leopards, and hyenas.
16. S. A. Parfitt et al., 'The Earliest Record of Human Activity in Northern Europe', *Nature* 438 (2005): 1008–12.
17. C. Stringer, *Homo britannicus: The Incredible Story of Human Life in Britain* (London: Allen Lane, 2006).
18. Well summarized in Dennell, *The Palaeolithic Settlement of Asia*.

19. L. B. Vishnyatsky, 'The Paleolithic of Central Asia', *Journal of World Prehistory* 13 (1999): 69–122; A. Markova, 'Fossil Rodents (Rodentia, Mammalia) from the Sel'Ungur Acheulean Cave Site (Kirghizstan)', *Acta Zoologica Cracoviensis* 35 (1992): 217–39.
20. Summarized in Dennell, *The Palaeolithic Settlement of Asia.*
21. M. D. Petraglia, 'The Lower Paleolithic of the Arabian Peninsula: Occupations, Adaptations, and Dispersals', *Journal of World Prehistory* 17 (2003): 141–79.
22. L. A. Schepartz, 'Upland Resources and the Early Palaeolithic Occupation of Southern China, Vietnam, Laos,Thailand and Burma', *World Archaeology* 32 (2000): 1–13.
23. C. Finlayson et al., 'Biogeography of Human Colonizations and Extinctions in the Pleistocene', *Memoirs Gibcemed* 1 (2000): 1–69.
24. Finlayson, *Neanderthals and Modern Humans.*
25. We used the terms sources and sinks which were well established in metapopulation biology, where source populations are those that produce a surplus of individuals each breeding season. These individuals can disperse to other source populations and especially to sink populations. Sink populations are marginal populations that are only maintained by immigrants as the local conditions are insufficient to maintain a resident population.
26. N. A. Drake et al., 'Ancient Watercourses and Biogeography of the Sahara Explain the Peopling of the Desert', *PNAS* 108 (2011): 458–62.
27. C. Finlayson, *Avian Survivors: The History and Biogeography of Palearctic Birds* (London: T. & A. D. Poyser, 2011).
28. I discussed *Homo heidelbergensis* at length in *The Humans Who Went Extinct* so I limit the discussion here to the relevance of

retaining *Homo erectus* as a more useful term. Finlayson, *Avian Survivors* and references therein.

29. A polytypic species is one that usually occupies a large geographical area and has significant regional variation. Populations at the extremes may appear quite different from each other but they are connected by others in between that interbreed and confirm that all populations, despite these differences, all belong to a single species.

CHAPTER 7

1. Finlayson, *The Humans Who Went Extinct*.
2. Barham and Mitchell (eds.), *The First Africans*. These authors provide a useful summary of the African evidence in the period between 430 and 70 kyr, which is the subject of this chapter.
3. Moments when population sizes declined severely, restricting their genetic diversity.
4. O. M. Pearson, 'Postcranial Remains and the Origin of Modern Humans', *Evolutionary Anthropology* 9 (2000): 229–47.
5. *The Humans Who Went Extinct*, 234 n. 46 lists references that confirm the endurance adaptation of the African *erectus* lineage.
6. At Omo Kibish in Ethiopia, dated to ~195 kyr. I. McDougall et al., 'Stratigraphic Placement and Age of Modern Humans from Kibish, Ethiopia', *Nature* 433 (2005): 733–6. But see also Preface.
7. Wolpoff, 'The Systematics of *Homo*'.
8. We can consider the different lineages, based on anatomical differences, as subspecies of *Homo sapiens*. The subspecies is a taxonomic category that refers to populations of a single

species that may be distinguished on the basis of differences of anatomy, colouration, voice, etc. They are different from each other but are able to interbreed and produce viable offspring. Using this convention, the African lineage would be *Homo sapiens sapiens*, the nominate subspecies. The Neanderthals, the Eurasian lineage, would be *Homo sapiens neanderthalensis* and *erectus* would be *Homo sapiens erectus*. Should we consider *heidelbergensis* to be sufficiently distinct from *erectus* to warrant subspecific status, then it would be *Homo sapiens heidelbergensis*. Similar criteria could be applied to other forms that are more contentious, e.g. *ergaster, rhodesiensis*, etc.

9. Generally referred to as Mode 3 technology, contrasting with Oldowan (Mode 1) and Acheulian (Mode 2). The previously prepared cores are referred to as Levallois cores. The technique allowed for greater control over the production of flakes; more could be extracted from a core, making the process more efficient than before. The method also permitted the making of stone-tipped spears, by hafting stone tools onto wooden shafts. These were the first composite tools known to have been made by *Homo*; Barham and Mitchell (eds.), *The First Africans*.
10. S. McBrearty, 'Down with the Revolution', in P. Mellars et al. (eds.), *Rethinking the Human Revolution* (Cambridge: McDonald Institute Monographs, 2007).
11. Robin Dunbar at the University of Liverpool has pioneered research into the importance of the brain in the context of the evolution of social groups. See, for example, R. Dunbar, *The Human Story: A New History of Mankind's Evolution* (London: Faber and Faber, 2004) and discussion in *The Humans Who Went Extinct*, 207.

12. Drake et al., 'Ancient Watercourses'; R. Tjallingii et al., 'Coherent High- and Low-Latitude Control of the Northwest African Hydrological Balance', *Nature Geoscience* 1 (2008): 670–5.
13. Finlayson, *The Humans Who Went Extinct*, 94–5.
14. Barham and Mitchell (eds.), *The First Africans*; V. Rots et al., 'Aspects of Tool Production, Use, and Hafting in Palaeolithic Assemblages from Northeast Africa', *Journal of Human Evolution* 60 (2011): 637–64.
15. *Ammotragus lervia.*
16. R. G. Klein and K. Scott, 'Re-Analysis of Faunal Assemblages from the Haua Fteah and Other Late Quaternary Archaeological Sites in Cyrenaica, Libya', *Journal of Archaeological Science* 13 (1986): 515–42.
17. A. Bouzouggar et al., '82,000-year-old Shell Beads from North Africa and Implications for the Origins of Modern Human Behaviour', *PNAS* 104 (2007): 9964–9.
18. The authors found ochre and that the shells were perforated, suggesting modern human behaviour which was linked to the exploitation of the coast. I disagree with the idea of a sudden emergence of modern behaviour, a concept which is, in any case, slippery and impossible to define.
19. C. Marean et al., 'Early Human Use of Marine Resources and Pigment in South Africa during the Middle Pleistocene', *Nature* 449 (2007): 905–9.
20. R. G. Klein, *The Human Career: Human Biological and Cultural Origins* (Chicago: University of Chicago Press, 1999).
21. H. Faure et al., 'The Coastal Oasis: Ice Age Springs on Emerged Continental Shelves', *Global and Planetary Change* 33 (2002): 47–56.

22. J. E. Yellen et al., 'The Archaeology of Aduma Middle Stone Age Sites in the Awash Valley, Ethiopia', *PaleoAnthropology* 10 (2005): 25–100.
23. Barham and Mitchell (eds.), *The First Africans.*

CHAPTER 8

1. Estimates vary but most cluster around 400 kyr, see *Humans Who Went Extinct*, 106–7.
2. The argument that the Neanderthals originated in Europe from Middle Pleistocene ancestors is prevalent in the literature, e.g. C. Stringer and C. Gamble, *In Search of the Neanderthals* (London: Thames and Hudson, 1993). It stems from the number of Neanderthal fossils that have been recovered in Europe but this may be a simple artefact of the sampling, western Europe having been subjected to intense excavation for over a century while sites in other parts of the Neanderthal range, in Siberia for example, have been less intensely studied.
3. C. Finlayson and J. S. Carrión, 'Rapid Ecological Turnover and Its Impact on Neanderthal and Other Human Populations', *Trends in Ecology and Evolution* 22 (2007): 213–22.
4. C. Finlayson, 'On the Importance of Coastal Areas in the Survival of Neanderthal Populations during the Late Pleistocene', *Quaternary Science Reviews* 27 (2008): 2246–52.
5. C. Finlayson et al., 'Late Survival of Neanderthals at the Southernmost Extreme of Europe', *Nature* 443 (2006): 850–3. These last dates have been disputed on various grounds but have been robustly defended so that they remain the oldest known Neanderthal dates: J. Zilhao and P. Pettit, 'On the New Dates for Gorham's Cave and the Late Survival of Iberian Neanderthals', *Before Farming* 3 (2006): 1–9; C. Finlayson

et al., 'Gorham's Cave, Gibraltar—The Persistence of a Neanderthal Population', *Quaternary International* 181 (2008): 64–71. Bradshaw and colleagues calculated the date when the Gorham's Cave Neanderthals disappeared to 31.5 kyr. They based their estimate on the assumption that the last dates, around 32 kyr, could not represent the last ones but a small surviving population instead: C. Bradshaw et al., 'Robust Estimates of Extinction Time in the Geological Record', *Quaternary Science Reviews* 33 (2012): 14–19.

6. R. Pinhasi et al., 'Revised Age of Late Neanderthal Occupation and the End of the Middle Paleolithic in the Northern Caucasus', *PNAS* 108 (2011): 8611–16; R. E. Wood et al., 'Radiocarbon Dating Casts Doubt on the Late Chronology of the Middle to Upper Palaeolithic Transition in Southern Iberia', *PNAS* doi/10.1073/pnas.1207656110. These authors have redated Neanderthal sites across Europe using a new method of cleaning bone prior to radiocarbon dating. They have concluded that most resampled dates were older than under the previous untreated dates. From this they have concluded that the Neanderthals went extinct rapidly across Europe around 40 kyr. The conclusions have several flaws. In the first place these dates around 40 kyr must represent a healthy Neanderthal population if it was leaving so much evidence across the continent so the Neanderthals must have disappeared after 40 kyr. Second, their method of sampling requires that sufficient collagen is preserved to allow the material to be dated. Collagen survives best in cold and dry environments. This is precisely where we would expect the Neanderthals to have disappeared first. In contrast, warm and wet conditions, where the Neanderthals survived late, are unsuitable for this

method of dating. So these results are biased towards the older dates. The late dates from Gorham's Cave were taken from charcoal and not bone. When this team subjected the Gorham's Cave charcoal to a similar cleaning treatment they found no difference in the treated and untreated dates: T. F. G. Higham et al., 'The Radiocarbon Chronology of Gorham's Cave', in R. N. E. Barton, C. B. Stringer, and J. C. Finlayson (eds.), *Neanderthals in Context: A Report of the 1995–1998 Excavations at Gorham's and Vanguard Caves, Gibraltar* (Oxford: Oxford University School of Archaeology Monograph 75, 2011).

7. T. Akazawa, K. Aochi, and O. Bar-Yosef, *Neandertals and Modern Humans in Western Asia* (New York: Plenum Press, 1998).
8. Finlayson and Carrión, 'Rapid Ecological Turnover and Its Impact on Neanderthal and Other Human Populations'.
9. Hence why the most recent dates from these regions have not crossed this threshold, see n. 6.
10. The Spanish Region of Andalucía and the British Territory of the Rock of Gibraltar combined; R. Jennings et al., 'Southern Iberia as a Refuge for the Last Neanderthal Populations', *Journal of Biogeography* 38 (2011): 1873–85.
11. Using a Geographical Information System (GIS) we constructed two climate maps drawing on present-day temperature and rainfall measurements from 338 weather stations across the study area.
12. C. B. Stringer et al., 'Neanderthal Exploitation of Marine Mammals in Gibraltar', *PNAS* 105 (2008): 14319–24.
13. M. Cortés-Sánchez et al., 'Earliest Known Use of Marine Resources by Neanderthals', *PLoS One* 6 (2011): e24026.

14. A number of authors argue that brain-selective nutrients, particularly the omega-3 fatty acid docosahexaenoic acid DHA and arachidonic acid AA, have been essential in the successful development of the human brain. They extrapolate this ontogenetic aspect to phylogenetics inferring that the nutrients were necessary for the encephalization of the human brain; see S. C. Cunnane and K. M. Stewart (eds.), *Human Brain Evolution: The Influence of Freshwater and Marine Food Resources* (Hoboken, NJ, Wiley-Blackwell, 2010).
15. K. Milton, 'Reply to S. C. Cunnane', *American Journal of Clinical Nutrition* 72 (2000): 1586–7.
16. Among Neanderthal sites, 88 per cent had wetland birds, 28 per cent had coastal birds, and another 28 per cent had marine birds present. Many sites seem to have been important wetlands as the average number of wetland bird species represented was ~6. C. Finlayson et al., 'The *Homo* Habitat Niche: Using the Avian Fossil Record to Depict Ecological Characteristics of Palaeolithic Eurasian Hominins', *Quaternary Science Reviews* 30 (2011): 1525–32.
17. P. Mellars, 'The Neanderthal Legacy: An Archaeological Perspective from Western Europe' (Princeton: Princeton University Press, 1996); Jennings et al., 'Southern Iberia'.
18. Finlayson, *Neanderthals and Modern Humans.*
19. Among Neanderthal sites 92 per cent incorporated rocky habitats with an average of ~11 rocky habitat bird species per site. Finlayson et al., 'The *Homo* Habitat Niche'.
20. F. J. Jiménez-Espejo et al., 'Environmental Conditions and Geomorphologic Changes during the Middle–Upper Paleolithic in the Southern Iberian Peninsula', *Geomorphology* 180–1 (2013): 205–16.

21. Studies of deep-sea marine cores, drilled from the seabed, provide useful information on climate. The different layers deposited in the seabed are picked up in the cores and can be dated. The chemical composition of the different layers informs about climatic conditions at the time of deposition.
22. Finlayson, *The Humans Who Went Extinct*; Finlayson, *Neanderthals and Modern Humans.*
23. M. I. Eren and S. J. Lycett, 'Why Levallois? A Morphometric Comparison of Experimental "Preferential" Levallois Flakes versus Debitage Flakes', *PLoS One* (2012): 10.1371/journal.pone.0029273.
24. See a recent article supporting this view: N. Boivin et al., 'Human Dispersal Across Diverse Environments of Asia during the Upper Pleistocene', *Quaternary International* (2013): <http://www.dx.doi.org/10.1016/j.quaint.2013.01.008>.
25. Pearson, 'Postcranial Remains'.
26. D. A. Raichlen et al., 'Calcaneus Length Determines Running Economy: Implications for Endurance Running Performance in Modern Humans and Neandertals', *Journal of Human Evolution* 60 (2011): 299–308.

CHAPTER 9

1. Out-of-Africa 2 to distinguish it from the earlier expansion of *erectus.* See, for example, C. Stringer and R. McKie, *African Exodus: The Origins of Modern Humanity* (London: Jonathan Cape, 1996) for the consensus view.
2. Finlayson, *The Humans Who Went Extinct.*
3. The Sahara was a land of lakes and large river networks, rich in fauna; Drake et al., 'Ancient Watercourses'.

4. For a useful recent discussion of the subject see N. Boivin et al., 'Human Dispersal across Diverse Environments of Asia during the Upper Pleistocene', *Quaternary International* (2013): <http://www.dx.doi.org/10.1016/j.quaint.2013.01.008>.
5. Stewart and Stringer, 'Human Evolution Out of Africa'.
6. See discussion of the way in which geographical range expansions and contractions take place in Finlayson, *The Humans Who Went Extinct*.
7. The enigmatic Denisovans have, mercifully, not been given a scientific name that would add further to the confusion. These people seem to have shared a common origin with *neanderthalensis* but did not contribute to the *neanderthalensis* gene flow into Eurasian *sapiens*. They did contribute between 4 and 6 per cent of their genetic material to the genomes of present-day Melanesians but not Australians. It is suggested that the Denisovans had an evolutionary history which was distinct from *neanderthalensis* and *sapiens*. J. Krause et al., 'The Complete Mitochondrial DNA Genome of an Unknown Hominin from Southern Siberia', *Nature* 464 (2010): 894–7; D. Reich et al., 'Genetic History of an Archaic Hominin Group from Denisova Cave in Siberia', *Nature* 468 (2010): 1053–60.
8. G. Barker, 'The "Human Revolution" in Lowland Tropical Southeast Asia: The Antiquity and Behavior of Anatomically Modern Humans at Niah Cave (Sarawak, Borneo)', *Journal of Human Evolution* 52 (2007): 243–61.
9. J. Bowler et al., 'New Ages for Human Occupation and Climatic Change at Lake Mungo, Australia', *Nature* 421 (2003): 837–40.
10. Finlayson et al., 'Biogeography of Human Colonizations'.

11. The Palaearctic is the biogeographical region that encompasses Eurasia, from the British Isles to Japan, southwards to include North Africa and the Middle East. The patterns described are in Finlayson, *Avian Survivors.*
12. Altitude replicates the climatic effects of latitude.
13. I use the term Himalayas loosely to mean all East Asian high mountain chains.
14. Finlayson et al., 'Biogeography of Human Colonizations'.
15. M. D. Petraglia and A. Alsharekh, 'The Middle Palaeolithic of Arabia: Implications for Modern Human Origins, Behaviours and Dispersals', *Antiquity* 77 (2003): 671–84.
16. L. V. Golovanova and V. B. Doronichev, 'The Middle Paleolithic of the Caucasus', *Journal of World Prehistory* 17 (2003): 71–140.
17. H. V. Nasab, 'Paleolithic Archaeology in Iran', *International Journal of Humanities* 18 (2011): 63–87.
18. The major difference between the stone-tool making technology on the northern and southern sides of the Caucasus Mountains indicates that the highest peaks marked a cultural boundary for *neanderthalensis.* O. Bar-Yosef et al., 'The Implications of the Middle–Upper Paleolithic Chronological Boundary in the Caucasus to Eurasian Prehistory', *Anthropologie* 44 (2006): 49–60.
19. R. Pinhasi et al., 'Revised Age of Late Neanderthal Occupation and the End of the Middle Paleolithic in the Northern Caucasus', *PNAS* 108 (2011): 8611–16.
20. Finlayson, *Neanderthals and Modern Humans.*
21. Finlayson, *The Humans Who Went Extinct.*
22. Finlayson, *The Humans Who Went Extinct.*
23. Finlayson, *Avian Survivors.*

24. Examples include many corvids, such as the choughs *Pyrrhocorax* and ravens, doves, eagles, creepers *Tichodroma*, and many others.
25. Finlayson, *The Humans Who Went Extinct*.
26. M. D. Petraglia et al., 'The Toba Volcanic Super-Eruption, Environmental Change, and Hominin Occupation History in India over the Last 140,000 Years', *Quaternary International* 258 (2012): 119–34.
27. Petraglia et al., 'The Toba Volcanic Super-Eruption', have argued for the survival of the Toba eruption by south-east Asian populations, citing the persistence of the Hobbit *Homo floresiensis* in the vicinity. The Hobbit may be an exception. Being small may have permitted its population to survive in isolation, by exploiting resources that would not have permitted the persistence of the larger *erectus*. In any case, our knowledge of *erectus* is limited, with a presence in the area around 100 kyr and a controversial claim down to 50 kyr.
28. Following from the Himalayan and Caucasus–Zagros launchpads, I extend the paradigm here to a different type of launchpad; in this case the Indo-Malayan launchpad would have worked through the opening of rainforest in cold and dry periods, linked to significant exposure of previously submerged land.
29. A. S. Mijares et al., 'New Evidence for a 67,000-Year-Old Human Presence at Callao Cave,Luzon, Philippines', *Journal of Human Evolution* 59 (2010): 123–32.
30. C. Finlayson, 'The Water Optimisation Hypothesis and the Human Occupation of the Mid-Latitude Belt in the Pleistocene', *Quaternary International* 300 (2013): 22–31.

31. Humid regime: rainfall range falls between 1,000 and 1,500 mm/year with 250–300 days of rain/year.
32. Semi-arid regime: rainfall range falls between 400 and 600 mm/year with 160–210 days of rain/year. Arid regime: rainfall range falls between 100 and 400 mm/year with 5–160 days of rain/year.
33. C. Lévi-Strauss, *Structural Anthropology* (New York: Basic Books).
34. Finlayson, 'The Water Optimisation Hypothesis'.
35. Three hundred and fifty-seven sites were identified as having had potential source populations. They covered all continents and the period 200–10 kyr. Several variables were recorded for each of the sites: whether it was a cave or an open air site; its proximity to lake, river, marsh, coast, or other wetland (within 5 km), that is sources of fresh water; and whether large terrestrial mammals, small terrestrial game, rocky habitat mammals, marine/freshwater mammals, or marine/freshwater small game were recorded. The last category included molluscs. In the case of coastline, the distance was estimated to the coastline when the site was occupied. The identified hot spots were southern Africa; the Levant; south-western Iberia; central Mediterranean; north-west Iberia/south-west France; the Circum-Alpine region; the Northern Caucasus/Black Sea; southern Siberia; southern Australia; coastal California; coastal Peru; and coastal Chile.
36. Inland wetlands were close to human occupation sites in 276 cases (77.3 per cent) and the remaining 81 sites (22.7 per cent) were all close to the coast. Coastal sites would have had access to sources of fresh water, particularly in cases of lowered sea levels when coastal oases would have dotted the emerged coastal shelf (Ch. 7).

37. G. Finlayson et al., 'Dynamics of a Thermo-Mediterranean Coastal Environment—the Coto Doñana National Park', *Quaternary Science Reviews* 27 (2008): 2145–52.

38. There is increasing evidence supporting the view that large areas that are today's Sahara Desert, Arabian Desert, and Thar Desert in India, were once wetter and populated. H. Groucutt and M. D. Petraglia, 'The Prehistory of the Arabian Peninsula: Deserts, Dispersals, and Demography', *Evolutionary Anthropology* 21 (2012): 113–25; H. V. A. James and M. D. Petraglia, 'Modern Human Origins and the Evolution of Behavior in the Later Pleistocene Record of South Asia', *Current Anthropology* 46 (2005): S3–27; M. Petraglia et al., 'Middle Paleolithic Occupation on a Marine Isotope Stage 5 Lakeshore in the Nefud Desert, Saudi Arabia', *Quaternary Science Reviews* 30 (2011): 1555–9; J. Rose, 'The Role of the Saharo-Arabian Arid Belt in the Modern Human Expansion', *Acta IV Congress Pen.* (2004): 57–67; P. van Peer et al., 'The Early to Middle Stone Age Transition and the Emergence of Modern Human Behaviour at site 8-B-11, Sai Island, Sudan', *Journal of Human Evolution* 45 (2003): 187–93.

39. M. I. Bird, D. Taylor, and C. Hunt, 'Palaeoenvironments of Insular Southeast Asia during the Last Glacial Period: A Savanna Corridor in Sundaland?', *Quaternary Science Reviews* 24 (2005): 2228–42.

40. These were groups concentrated around the central Mediterranean and the Pacific Rim and they were also linked to freshwater or coastal locations.

CHAPTER 10

1. J. Flood, *Archaeology of the Dreamtime: The Story of Prehistoric Australia and Its People* (Marleston, SA: J. B. Publishing, 2004).
2. L. Beaufort et al., 'Biomass Burning and Oceanic Primary Production Estimates in the Sulu Sea Area over the last 380 kyr and the East Asian Monsoon Dynamics', *Mar. Geol.* 201 (2003): 53–65; G. Anshari et al., 'Environmental Change and Peatland Forest Dynamics in the Lake Sentarum Area, West Kalimantan, Indonesia', *J. Quat. Sci.* 19 (2004): 637–55; Finlayson, *The Humans Who Went Extinct*, 87.
3. Rapid Inland Expansion view: J. B. Birdsell, 'The Recalibration of a Paradigm for the First Peopling of Greater Australia', in J. Allen, J. Golson, and R. Jones (eds.), *Sunda and Sahul: Prehistoric Studies in South East Asia, Melanesia and Australia* (London: Academic Press, 1977), 113–67. Coastal and Riverine Expansion view: S. Bowdler, 'The Coastal Colonization of Australia', in Allen et al. (eds.), *Sunda and Sahul*, 205–46; Entry via Well-Watered Areas view: D. R. Horton, 'Water and Woodland: The Peopling of Australia', *Australian Institute of Aboriginal Studies Newsletter* 16 (1981): 21–7.
4. Smith first reported dates of ~22 kyr but has now extended the date of first occupation to ~35 kyr. M. A. Smith, 'Pleistocene Occupation in Arid Central Australia', *Nature* 328 (1987): 710–11; M. A. Smith, 'Characterizing Late Pleistocene and Holocene Stone Artefact Assemblages from Puritjarra Rock Shelter: A Long Sequence from the Australian Desert', *Records of the Australian Museum* 58 (2006): 371–410.
5. S. Bowdler, '*Homo sapiens* in Southeast Asia and the Antipodes: Archaeological Versus Biological Interpretations', in T. Akazawa,

K. Aoki, and T. Kimura (eds.), *The Evolution and Dispersal of Modern Humans in Asia* (Tokyo: Hokusen-sha, 1992).

6. Finlayson, *The Humans Who Went Extinct*; Boivin et al., 'Human Dispersal Across Diverse Environments of Asia'.
7. P. Veth, 'Islands in the Interior: A Model for the Colonization of Australia's Arid Zone', *Archaeology in Oceania* 24 (1989): 81–92.
8. Kulpi Mara rock shelter: P. B. Thorley, 'Pleistocene Settlement in the Australian Arid Zone: Occupation of an Inland Riverine Landscape in the Central Australian Ranges', *Antiquity* 72 (1998): 34–45. Serpent's Glen rock shelter: S. O'Connor et al., 'Serpent's Glen Rockshelter: Report of the First Pleistocene-Aged Occupation Sequence from the Western Desert', *Australian Archaeology* 46 (1998): 12–22.
9. M. A. Smith, 'The Case for a Resident Human Population in the Central Australian Ranges during Full Glacial Aridity', *Archaeology in Oceania* 24 (1989): 93–105.
10. Smith, 'Characterizing Late Pleistocene and Holocene Stone Artefact Assemblages'.
11. P. Clarke, *Where the Ancestors Walked* (Crows Nest, NSW: Allen and Unwin, 2003).
12. P. Hiscock, *Archaeology of Ancient Australia* (London, Routledge, 2008).
13. The crowning achievement of *Homo sapiens*: the ability of a water-dependent hominid to survive and succeed in some of the most water-deficient environments on Earth.
14. R. Tonkinson, *The Mardu Aborigines: Living the Dream in Australia's Desert* (Mason, O.: Cengage Learning, 2008).
15. Lerp: a sweet secretion left by scale insects on leaves.
16. Following R. A. Gould, 'The Anthropology of Human Residues', *American Anthropologist* 80 (1978): 815–35.

17. The Mardu, who were strong advocates of a mixed diet, were adequately nourished, had well-balanced diets, and showed no evidence of vitamin or protein deficiency.
18. The Mardu seasons: Dulbarra, akin to spring; Yalijarra, the hottest part of the year; and Wandajarra, the coldest time of the year. Each season was identified with particular resources that became available.
19. A. Radcliffe-Brown, 'Notes on the Social Organization of Australian Tribes', *Journal of the Royal Anthropological Institute of Great Britain and Ireland* 48 (1918): 222–53.
20. The account of the Murray–Darling people is derived largely from C. Pardoe, 'The Cemetery as Symbol: The Distribution of Prehistoric Aboriginal Burial Grounds in southeastern Australia', *Archaeology in Oceania* 23 (1988): 1–16; C. Pardoe, 'Riverine, Biological and Cultural Evolution in Southeastern Australia', *Antiquity* 69 (1995): 696–713.
21. Skull deformation is the practice of tightly binding a baby's head with cloth, elongating the skull. In the case of the Murray–Darling people, deformation seems to have been done by the adults applying constant pressure with thumbs and palms to the foreheads of the newborn children. Tooth avulsion is the practice of knocking out specific teeth during a rite of passage from adolescence to adulthood.

CHAPTER 11

1. H. M. Monroe, 'Australia: The Land Where Time Began. A Biography of the Australian Continent: Puritjarra Cave Rock Shelter', <http://www.austhrutime.com/puritjarra_Cave_Rock_Shelter.htm>.

2. An ancient lake in the heart of Middle Earth alludes to Lake Chad where Toumaï lived. We do not know if this is where the story started but it would certainly have been a wetland within the core area of Middle Earth. The second story refers to the start of agriculture in the Fertile Crescent, possibly on the banks of the Tigris and Euphrates rivers in ancient Mesopotamia, also in the heart of Middle Earth.
3. See, for example, I. Herbinger et al., 'Territory Characteristics among Three Neighboring Chimpanzee Communities in the Taï National Park, Côte d'Ivoire', *International Journal of Primatology* 22 (2001): 143–67.
4. Stringer et al., 'Neanderthal Exploitation of Marine Mammals in Gibraltar'.
5. C. Shipton et al., 'Variation in Lithic Technological Strategies among the Neanderthals of Gibraltar', *PLoS One* 8 (2013): e65185.
6. The Younger Dryas; Finlayson, *The Humans Who Went Extinct*, 197.
7. By people staying in wet upland refuges during arid periods and spreading along lowland corridors, associated with water, during wet periods; P. Veth, 'Islands in the Interior: A Model for the Colonization of Australia's Arid Zone', *Archaeology in Oceania* 24 (1989): 81–92; M. A. Smith, 'Biogeography, Human Ecology and Prehistory in the Sandridge Deserts', *Australian Archaeology* 37 (1993): 35–50.
8. P. Bellwood, *First Farmers: The Origins of Agricultural Societies* (Oxford, Blackwell, 2005); T. Denham and P. White, *The Emergence of Agriculture: A Global View* (London: Routledge, 2007); J. Diamond, *Guns, Germs and Steel: A Short History of Everybody for the Last 13,000 Years* (London, Jonathan Cape, 1997).

9. Canadian biologist Doug Larson has advocated the close links between the modern urban human habitat and the rocky places that we frequented in our past; D. Larson et al., *The Urban Cliff Revolution: New Findings on the Origins and Evolution of Human Habitats* (Markham, Ont.: Fitzhenry and Whiteside, 2004).

INDEX